STUDIES IN MULTIDISCIPLINARITY VOLUME 6

Reviving the Living

Meaning Making in Living Systems

STUDIES IN MULTIDISCIPLINARITY

On the cover:
Imaginary Garden by Tamar Neuman

STUDIES IN MULTIDISCIPLINARITY VOLUME 6

Reviving the Living

Meaning Making in Living Systems

Yair Neuman
Ben-Gurion University of the Negev
Beer-Sheva, Israel

ELSEVIER

Amsterdam – Boston – Heidelberg – London – New York – Oxford
Paris – San Diego – San Francisco – Singapore – Sydney – Tokyo

Elsevier
Radarweg 29, PO Box 211, 1000 AE Amsterdam, The Netherlands

First edition 2008

Library of Congress Cataloging-in-Publication Data
A catalog record for this book is available from the Library of Congress

British Library Cataloguing in Publication Data
A catalogue record for this book is available from the British Library

ISBN: 978-0-444-53033-2
ISSN: 1571-0831

For information on all Elsevier publications
visit our website at books.elsevier.com

Transferred to Digital Printing 2011

Series Dedication

Studies in Multidisciplinarity is dedicated to the memory of Ray Paton.

Sure, he that made us with such large discourse,
Looking before and after, gave us not
That capability and god-like reason
To fust in us unused.
– William Shakespeare, Hamlet

As always, to my beloved children Yiftach, Yaara, and Tamar

Acknowledgments

Rabbi Moshe Ben-Maimon (1135–1204), known as the Maimonides, was an eminent Jewish scholar, physician, and philosopher. In his treatise *Mishneh Torah* he wrote an insightful statement about the relation between a person and his teacher:

> Just as a person is commanded to honor and revere his father, so it is his duty to honor and revere his teacher, even more than his father; for his father has secured him life in this world, while the teacher who has taught him wisdom secures for him life in the future world.

The Hebrew word for a teacher is far remote from the English translation. The etymological source of "teacher" tells us that it was used to denote the slave who escorted children to school. In contrast, the Hebrew term denotes the activity of pointing at the right direction. According to the Jewish sense, a "teacher" can only point at the way. The context of my acknowledgements is not the same context of the Maimonides teaching. However, the Maimonides statement draws an important analogy between being a good "teacher" and being a good parent. This analogy is important for understanding the meaning of learning and the role of significant others in our personal development. It is my pleasure to thank two significant others, Irun Cohen for teaching me the wisdom of the immune system and Peter Harries-Jones for teaching me the wisdom of the social systems.

I would also like to thank Steven Rosen for "recursive dialogues", Meni Neuman for a constructive reading, my university rector Jimmy Weinblat for supporting the publication of the book, Jeanette Bopry for her excellent editorial work, Michael Weinstock for correcting my "Hebrew English" to "American English", and Mouton de Gruyter for their permission to reprint copyright materials.

Finally, I had the pleasure to work with two professional and friendly editors who made the best efforts to support this project. I would like to thank them both warmly: Laura A. McNamara, the academic editor and Elsevier's editor Anne Russum.

Dedication

One day my Grandfather called me on the phone and wished me a happy birthday. I was pleased that the old man remembered the birthday of his first grandson, and I was interested in the mnemonic tactic he used to recall the date. "Very simple", my Grandfather explained "I write all the birthdays' dates of my children, my grandchildren, and my Great-grandsons in my daily prayer book".

For an unknown reason at that time, this story excited me. I shared this excitement with a narrow-minded friend who dismissed my Grandfather's explanation as simply reflecting the mnemonic technique of an old man. However, for me, my Grandfather's mnemonic technique was full of meaning. In retrospective, I realize that the association between the Holy words of the prayer, between the most abstract and transcendental concept, the one of God, and the most concrete and localized event of an individual's birthday is something that perfectly characterizes my Grandfather and my Grandmother. Not a detached abstract thinking neither meaningless concrete nor particular activities but life *in between*. I have realized then that beyond the generation gap, the cultural gap, the age gap, the educational gap, or any other gap between me and my grandparents, each of us is trying in his own way to live the logic of in between.

This book is about meaning making in living systems, a novel perspective for understanding the realm of the living. However, underneath the surface the book is about the logic of *in between* and the way living beings manage their way in the world by orchestrating a delicate balance between order and disorder, past and present, the abstract and the concrete, and the static and the dynamic. This underlying logic of the book provides me with a good excuse to dedicate my second book to my Grandfather Meir Lifshitz and to my Grandmother Bracha (Berta) Lifshitz, two good and honest people who taught me, in their own way, a lesson about the meaning of life beyond its molecular level.

Preface

> Among the unpaved ways one is mine
>
> —Vladimir Vissotsky, "*Shattering*"

Opening a book is always a challenge especially if it appears under the title *Studies in Multidisciplinarity*. Disciplines are organized around common themes, concepts, problems, shared enemies, and other features that give them a sense of coherence. This disciplinary ground allows an author of a disciplinary book to address his audience without "foreplay" and to get directly to the point. This is the reason why I usually find disciplinary books to be so boring. Both in lovemaking and in book writing, foreplay is of indispensable value.

An author who addresses different audiences and challenges them from an interdisciplinary perspective is facing a problem: How to introduce your book while having no simple common ground with your audience? I was struggling with the question of how to open this book, then one day I watched, with my kids, Walt Disney's version of *Alice's Adventures in Wonderland*. In one of the scenes Alice arrives to the tea party of the Mad Hatter. In the party she meets the Mad Hatter, the March Hare, and the Dormouse. She tries to share with them the adventures she has had but does not know where to begin. Ah! This was exactly my situation and I hushed my kids in order to learn a lesson from the Mad Hatter. The advice Alice receives from the Mad Hatter and the March Hare is illuminating and impressive in its simplicity: *Start at the beginning and stop at the end.* Apparently, this advice should inspire any rational author. However, Lewis Carroll's characters are not the best models for rational thinking. What is wrong with the Mad Hatter's appealing and illuminating advice? The answer is that, as creatures with consciousness and memory, we have no simple beginning ready to hand, no Euclidean point from which we may start. For any beginning there is a previous and/or alternative beginning from which we may begin to tell our story. Indeed, as all other living creatures, we are born, live, and end our lives immersed in intricate webs in which a straight path from "The Beginning" to "The End" is seldom observed. In this context, the Mad Hatter's advice, with all of its apparent rationality, is the advice of a mad man who is seeking something that does not exist. Therefore, opening a story, a lecture, or an academic book can never start at "The Beginning" unless the Big Bang or the Book of Genesis has clear relevance to our story.

Where should I start? I would like to start with a feeling of dissatisfaction, which may be wrongly interpreted by some people as a religious or more accurately, a scientific heresy. This feeling of dissatisfaction concerns our understanding of living systems, and the way this understanding, in a very deep sense, banishes organisms from their unique status as living systems.

Our knowledge of living systems celebrated its alleged victory through downward reductionism. Let us take two disciplines as an example: biology and linguistics. Concerning our own species, both in biology and linguistics, we improved our ability to break the system into small component parts and to achieve a better and better understanding of those parts and their organization. The Genome project has increased our knowledge of the way the genetic "letters" are organized and Chomsky's theory of grammar has made a similar contribution with regard to the way in which the basic components of a sentence are organized.

The advancement of knowledge both in biology and linguistics cannot be denied. However, living systems are more than stubbles of cells (or genes) just as language is more than a collection of linguistic signs. Each expresses the gestalt property of a whole which is different from a collection of its parts; a whole that exists only as long as it is constituted by its interacting parts, by interacting with itself, and by its interaction with its environment. No language exists without a community of language users arguing, asking, joking, explaining, tempting, and constructing a shared reality. Along the same line, our genes are meaningless if they are taken out of the context of a living organism.

Context, as its Latin etymology teaches us, comes from *contexere* "to weave together". Understanding context is understanding the way things are woven together in a network (*textere*). Without understanding the way things are woven together, we are left with a fragmented and mechanistic conception of the realm we would like to understand. This book is a study in biological weaving; it aims to revive the study of living creatures that have turned, under the influence of reductionism, into a dead stubble of genes. This is the reason why I titled the book *Reviving the Living.*

The holistic nature of living systems is a fact that cannot be denied (Noble, 2006) although it is definitely a fact that can be oppressed. It is important to recognize the conclusion that is necessarily derived from this observation. If living systems are gestalt-like wholes that are constituted through micro-level interactions then *ipso facto* reductionism is extremely limited in helping us to understand them. This conclusion sets a clear barrier to our understanding of living systems. Digging more and more deeply into the components of living systems would not help us to understand the meaning of their behavior. Like children who have successfully uncovered the components of a mechanical toy, we are now facing the challenge of understanding the working whole. This conclusion is valid both in biology and natural language, two fields that will occupy my attention in this book.

The reader should not mistake the above conclusion for a naïve holistic alternative that aims to dismiss the proven benefits of reductionism or to replace scientific rigor with general and vague non-scientific terminology. This is clearly not the alternative that I am seeking and the reader should avoid the straw man fallacy while critically judging my thesis.

Now, let us assume that living systems are interactive wholes and that we would like to understand them as interactive wholes. Is there a non-reductionist alternative? The reader familiar with the "complexity sciences" may immediately think he or she has the answer. Indeed, the idea of macrostructures that emerge out of micro-level interactions was found illuminating in certain respects. However, I would like to address the challenge of a non-reductionist alternative from a different perspective. To introduce this alternative, I would like to quote an insightful excerpt from *Art and Answerability*, a treatise written by the Russian polymath Mikhail Bakhtin in 1919.

Bakhtin is unfamiliar to biologists and hardly well known to philosophers, but just listen to what he has to say:

> A whole is called "mechanical" when its constituent elements are united only in space and time by some external connection and are not imbued with the internal unity of *meaning*. (Bakhtin, 1990, p. 1; emphasis mine)

Bakhtin is making an important statement, which is the cornerstone of this manuscript. The internal unity of a "non-mechanical" whole, a living being, is achieved through the *internal unity of meaning*. What does he mean by *meaning* and is meaning a key concept for our understanding of living systems as wholes? The second question is still an open question and I hope to make a case for the positive answer in this book.

Excluding several rare cases such as Anton Markos' (2002) *Readers of the Book of Life*, Hoffmeyer's (1996) *Signs of Meaning in the Universe*, or Marcello Barbieri's (2002) *The organic codes*, the concept of meaning has not been directly faced in the scientific literature dealing with theoretical biology. Even in these important texts meaning, in the sense in which it is discussed in this manuscript, is not the main Organizing concept.

Not only meaning is not a major organizing concept in biological research but meaning has been *excluded* from our understanding of biological systems by mainstream biology. Modern biology with its enthusiasm for the genome and its "information" content, whatever that means, has adopted an information processing approach. The reader may find many papers dealing with *Bio-Informatics* but what about *Bio-Meaning*? Meaning has been excluded from information theory by Shannon, its godfather, so we are left only with information. Unfortunately, meaning has been left to the philosophers.

This book aims to present the idea that living systems are meaning-making systems, to develop a meaning-making perspective on living systems, and to illustrate the fruitfulness of this perspective by using concrete examples. Examples are taken from several domains: genetics, the immune system, and natural language.

At this point, a qualification should be added. I do not consider *meaning* to be the new philosophers' stone that will bring us to the light of ultimate understanding. I oppose any kind of theoretical reductionism that aims to reduce the multidimensional nature of living systems to a single explanatory dimension, whether grammar in natural language or genes in biology. This is the reason why I do not, at this time, present a *theory* of meaning making, but just a new perspective, which is only one way of approaching the realm of the living. I do believe, however, that approaching living systems as meaning-making systems may change the way we study living systems and provide us with a different perspective which is scientifically rigorous, while at the same time faithfully representing the holistic, interactive, and sign-mediated (semiotic) nature of living systems.

At this point, I would like to use again the sexual metaphor of foreplay. Why foreplay again? Meaning is a concept that appears in several disciplines such as linguistics, semiotics, and psychology and clarifying the meaning of *meaning* and its relevance for living systems necessarily involves a multidisciplinary perspective. In this context, getting to the point is missing the point. Foreplay is therefore necessary. Readers will have to trust me while guiding them in the intricate web in which meaning is woven and bringing them finally to a meaning-making perspective on living systems. Patience is expected from the reader. We should remember that learning involves trust and participatory activity. As suggested by Deleuze (1994):

> We learn nothing from those who say: "Do as I do". Our only teachers are those who tell us to "do with me", and are able to emit signs to be developed in heterogeneity rather than propose gestures for us to reproduce. (p. 23)

"Do with me" is my only request as we enter into the realm of the living.

Prologue

1. The Audience

Several words should be said about the audience, the style, and the plan of the book. First, the major thesis presented in this book relies on ideas, concepts, problems, techniques, and examples from a variety of domains that converge toward the same conclusion. Readers may find a wealth of knowledge on biology and semiotics, the field that studies signs and signification. However, they will be also introduced to ideas from other fields such as immunology, philosophy, physics, and mathematics. Indeed some of the ideas presented in this book have been published in journals from various disciplines such as *Perspectives in Biology and Medicine*, *Progress in Biophysics and Molecular Biology*, *Semiotica*, *Rivisita di Biologia/Biology Forum*, and *Information Sciences*. However, the nature of the book is such that there is no expectancy that readers be expert in one or more of these domains but just educated readers who are willing to accept the challenge of approaching living systems in a novel way. Frankly, I do not consider myself to be anything other than, hopefully, an educated reader in these fields.

2. The Style

I have made an effort to present ideas in an instructive way and to design a book which is to a large extent self-contained. An inevitable result is a certain level of redundancy and simplicity for the expert, which is compensated by an instructive value for the educated reader. However, in certain places the book is theoretically dense and the non-linear nature of the book does not make things easier. Readers should be patient of the cognitive load that is expected of them in these sections.

Now, we get to the next issue, which is the style of the book. Readers who are familiar with my previous book (Neuman, 2003a) may be prepared for my style of writing, which is informal, reflective, and in certain places provocative and rather *politically incorrect*. There are two reasons for using this style. The first reason is that this style is the style I like when I read other people's books. The second reason for using this style is that I truly believe that, in contrast to an informative academic paper, a book must be enjoyed and thought challenging in order to justify the reader's (and the author's) time and exertion.

3. Intellectual Necrophilia or Why isn't This a Typical Philosophical Essay?

There is a long scholarly tradition in the humanities that those who are trying to offer a new solution to an old philosophical problem will critically review the history of the problem. I have a great respect for this tradition and followed it in my previous book. However, I realized quite recently that in many cases this tradition pathologically slips into a kind of *intellectual necrophilia.*

Necrophilia is defined in medical dictionaries as "A morbid [sexual] fondness for being in the presence of dead bodies". What I describe as an intellectual necrophilia is a state in which the scholar invests his entire libido and passion in the dead corpus of past scholars/ideas. Maybe this is the reason why I find philosophical journals to be so boring and why our scientific knowledge progresses with indifference to the work done in philosophy. Science looks forward while many philosophers look backward. Indeed, the past may teach us a lesson and the work of ancient philosophers may have great relevance for our understanding of current events, but at what point does a healthy interest in the past turns into perversion?

I believe that a crucial term for understanding the source of this intellectual pathology is *dialogue*. As the provocative Lacanian psychoanalyst and philosopher Slavoi Žižek once argued, beyond rhetoric, the great philosophers were not truly interested in dialogues (Žižek and Daly, 2004) but in providing their own unique perspectives on the world. This argument is dubious but it raises an important issue. A dialogue assumes a response and a response cannot be gained from a dead corpus. A response can be gained from nature through experimentation or from other people through dialogue. This is why Socrates considered the market place—the *Agora*—as the most appropriate place for actualizing *philosophia*—the love of wisdom; this is the reason why Talmudic scholars discussing ancient texts never consider their dialogues as subordinate to the dead corpus, but *vice versa*. In both cases living beings, with their concrete and daily problems in all their complexity, are the source of philosophizing.

Again, I do not dismiss the importance of a philosophical/conceptual analysis, but I would like to emphasize that in order to be relevant, a philosophical analysis must respond to current challenges and present a *prospective vision*. Following this suggestion my use of, and reference to, theoretical terms and problems will be guided by their clear relevance to current problems and the progress of knowledge, and not by a complete survey of past philosophical works.

4. The Approach: Multidisciplinarity and the Importance of Nomads-by-Choice

Multidisplinarity might have a bad reputation due to the work of some charlatans who flutter between disciplines while providing no interesting ideas, hypotheses, or insights. The reader is probably familiar with books and papers that produce the salad of quantum mechanics, brain science, and mysticism or with those that mix the second law of thermodynamics with creationism. When I encounter those who produce the quantum salad of mind and matter, I am usually tempted to ask them questions such as: What is the equivalence to a complex number in the mental realm? Is there a Hilbert Space of the mind? Or what is the equivalent of vector projection in the soul? The answers to these questions usually take the form of general, too general, abstract statements that go beyond the mathematical formulation and the insights that characterize the theory of quantum mechanics. These "inter" or "multi" disciplinary writers do not serve as evidence against the kind of a multidisciplinary approach that I use in this book. As Nietzsche once commented, the followers of a system cannot be used as evidence against it. So, why multidisciplinarity?

Unfortunately, the enormous amount of knowledge human beings have acquired has not been accompanied by a corresponding increase in brain function or by a proportional increase in their moral behavior. Our brains remain the same as the brains of our ancestors who held only a negligible fraction of the knowledge we hold today. The asymmetric development of scientific knowledge and brain functions makes it impossible for a single person to hold a firm grasp on several domains of inquiry. We simply do not have the cognitive resources to cope with this amount of information in a meaningful way.

Let me illustrate the difficulty by using enzymes. Enzymes are protein molecules that have a significant role in metabolic processes. If you are a researcher in genetics, enzymes are crucial for your understanding, for example, of the transformation from DNA to proteins. In this case you should be familiar with the literature on genetics as well as with the literature on enzymes, which is quite a burden. Enter into *PubMed* the search term *Enzymes* and you will get 1,503,900 hits! That is one million five hundred three thousand and nine hundred items. Although the irrelevance of some papers, as well as the redundancy of findings, concepts, and methods certainly reduces the cognitive load of this list, the number of items relevant to our subject matter is still too big to handle for our limited brain. However, things are even more complicated than that since there are many other fields that may be relevant for understanding the genetic system. We will not discuss them here, but I am saying that taking them into consideration will definitely increase your cognitive load.

A natural solution to cognitive load is to focus our attention only on a small portion of available stimuli. This solution is evident in the partition of science into smaller and smaller domains of expertise. This rather natural move has its price: As we turn into experts we lose a sense of the system as a whole. This price is clearly evident in modern medicine. The modern physician sometimes encounters simple medical problems that can be easily solved through modern technology. For example, diagnosed early, an infected appendix may be (relatively) easily removed by a surgical intervention. In other cases things are not so simple. The body is a complex in which different systems interact. In many cases we encounter problems that result from the interactions between these systems and when this happens our expertise might be an obstacle rather than advantage. Let me give you an example.

The literature on anesthesia suggests that in some cases during surgery patients might experience awareness under anesthesia even though they are supposed to be unconscious. According to this argument, the patient is mentally awakened during the surgery but with his/her muscles paralyzed. As a result of this horrible situation he or she might suffer from Post Traumatic Stress Disorder (PTSD) (Macleod and Maycock, 1992). Assuming that this argument is empirically grounded, what is the source of this trauma? A failure of the anesthesia? One possible explanation concerns the immune system. We should remember that the immune system functions as a diffuse sense organ. Is it possible that the trauma discussed in the anesthesiology literature is mediated by the immune system, which is activated during the surgery?

In one of my papers (Neuman, 2004a), I raised the hypothesis that unconscious pain experienced during general anesthesia may be mediated by the immune system. This hypothesis may have relevance for anesthesiologists but anesthesiologists usually do not read papers by immunologists, immunologists are usually unfamiliar with the domain of anesthesiology, and pain researchers seldom acquire expertise in immunology or anesthesiology. Therefore, interesting things that happen between systems do not get attention from members of the fragmented disciplines. Let me give you another example.

Postoperative intestinal adhesion is a common and painful problem for many people who experience abdominal surgery. As a result of this surgical intervention, adhesion is expected to appear in up to 90% of patients (Koessi *et al.*, 2003). Any physician who observed the pain resulting from intestinal adhesion probably sought for a solution, however, surprising as it may sound, there is no commonly agreed and scientifically tested procedure for preventing adhesion. Notice that the intestine is a relatively simple device: just a tube. Think about our ignorance concerning more complex organs such as the liver or the brain. Again, the problem is that biological adhesion is a subject matter, which is investigated in a number of disciplines from physics to zoology. To understand postoperative intestinal adhesion one has to understand the underlying physics of adhesion and the way microscopic elements interact to stick things together, the immunological process that takes place as a response to the damage of the tissues that result from surgical intervention and might mediate the adhesion, the biological mechanisms that may correct the adhesion, and so on. One is unlikely to find a human mind that is both an expert and a talented integrator of these fields; the bottom line is that a preventive solution has not yet been found to this disturbing and allegedly simple problem.

Given the amount of knowledge and our limited brain capacity, is there a way to bridge the gap? An interesting suggestion comes from a film director. The famous Polish film director Krzysztof Kieslowski said once that in order to become a talented film director one should learn psychoanalysis, theology, philosophy, and many other fields relevant for understanding the human experience. However, since we cannot really be experts in all these domains we are left with *intuition* only. Is this the solution to the problem I previously discussed? Intuition?

There is no doubt that intuition has a crucial role in our life as human beings and as researchers. However, like a mistress, and I use a metaphor once used by Françoise Jacob, it is something that exists but cannot be publicly admitted. Let me explain why by using Bergson's understanding of intuition. To quote:

> An absolute can only be given in an intuition, while all the rest has to do with analysis. We call intuition here the sympathy by which one is transported into the interior of an object in order to coincide with what there is unique and consequently inexpressible in it. Analysis, on the contrary is the operation which reduces the object to elements already known. (Bergson, 1946, p. 161)

Bergson presents analysis and intuition as diametrically opposed. Intuition involves sympathy and coincidence with the object. In contrast, analysis involves detachment from the object. Since science is usually portrayed as both a reductionists and an analytic profession, intuition is not the coin of the current scientific realm. A biologist is expected neither to feel sympathy toward the object under inquiry, nor to grasp it in holistic manner. Can you imagine a biologist transported into the interior of an *Escherichia Coli* in order to coincide with what there is unique and consequently inexpressible in it? Therefore, intuition is clearly indispensable for the working scientist, but for communicative rhetorical use it has no value.

So, what solution can we offer to the need to understand interactive wholes while having no ability to master all disciplinary knowledge?

A solution does not yet exist, but one helpful heuristic for achieving some answers is to recognize the importance of *nomadic researchers* who may wander between the fields with the aim of providing an integrative and novel perspective from the bird's-eye view. This is not my original proposal. Benoit Mandelbrot, the father of fractal geometry, once said:

> The rare scholars who are nomads-by-choice are essential to the intellectual welfare of the settled disciplines.

A man like Gregory Bateson is an excellent example of a nomad-by-choice who was not seeking answers within an established discipline but wandered between disciplines and domains to provide answers to questions that he saw as crucial to human understanding and survival. Indeed as a nomad-by-choice, I would like to propagate the importance of a multidisciplinary perspective in the same sense.

5. Check Your Schemes at the Entrance!

People, like other living systems, are creatures of habit. However, habits/schemes involve the danger of ignoring novelties and reading new texts through old glasses. This is the reason why I ask the reader to pay close attention to the novelties of the argument presented in this book. More specifically the manuscript developed a thesis which is in sharp contrast with information processing approaches to biology, which ignore meaning as a crucial concept for understanding living systems. I will also criticize the idea that living systems are Turing machine style and the ignorance of interaction as another constituting aspect of living systems. You will find me criticizing Chomsky, Turing, Shannon, Saussure, and many other scholars. On the other hand, you will find me following the tradition of several great scholars (Bakhtin, Bateson, Volosinov, Piaget, Polanyi, and Peirce) who emphasized the dynamic, non-representational, interactive, and "semiotic" nature of cognition as a "meaning-making" process which is created "in-between" levels of organization. None of these scholars are vitalists and my discussion clearly rejects neo-vitalism. I present the processes through which meaning is constructed "in-between" scales of analysis and as always embedded in a specific context of interaction. It is therefore highly important to grasp the significance of "context" as a constituting concept in the process of meaning making. In this context, I introduce terminology which is probably new to many readers: polysemy, dual coding, boundary conditions, transgradience, mesoscopic, and so on. The terminology will be introduced step-by-step in an onion-like and reflective fashion like the topological structure known as the Klein bottle, which is another concept introduced later. Therefore, I repeat my advice, "Check your schemes at the entrance", and do not read me through the glasses of the known.

6. The Plan

The book is divided into several parts. The first part is entitled: "How Low Can You Go?" In this part, I present the idea of scientific reductionism and point at its limitations by drawing on two fields: genetics and immunology. The second part of the book presents a meaning-making perspective while resonating with three fields of language study: syntax, semantics, and pragmatics. The third part of the book aims to discuss several aspects of meaning making from a radical standpoint. For example, the polysemy of the sign is discussed in terms of a *superposition*. The fourth and concluding part of the book is an attempt to reflect on meaning making from a highly abstract and poetic perspective.

After briefly presenting each field there are chapters entitled: "A Point for Thought". In these chapters I present some radical ideas on how to approach certain problems within these fields from a new perspective of meaning making. Therefore, the book is not organized in a linear way and some relatively complex ideas regarding meaning making in living systems appear before the perspective has been systematically presented. The reader should not be discouraged from this style of writing and trust himself to grasp at least the radical *Geist* that appears in these sections.

7. Cat-logues

My previous book included imaginary dialogues with my cat, Bamba. The dialogues were used to reflect on some themes within the book from a humorous and critical perspective. The reviewers of the book and its readers found great joy in these cat-logues, so I have decided to use them in the current book as well. In retrospective it seems only natural that a book about living systems should include the perspective of a non-human organism.

8. What Are Books Good for?

Mark Twain was once asked by a peasant to explain what books are good for. Twain replied by saying something like this: Thick books are excellent for stabilizing shaking chairs, books covered with leather are excellent for sharpening shaving razors, and hard-covered books are excellent for striking the heads of those stupid enough to ask what books are good for.

Twain's witty response was appropriate for an era in which books were relatively rare and people relatively ignorant. Today, publishing a book for an educated audience needs a better justification. I do hope that the readers of this book will find at least one thing that this book is good for besides for striking the heads of those stupid enough to ask what books are good for.

Table of Contents

Series Dedication v

Dedication vi

Acknowledgments vii

Dedication ix

Preface xi

Prologue xv

PART 1. HOW LOW CAN YOU GO? REDUCTIONISM AND ITS LIMITATIONS

CHAPTER 1: WHAT IS REDUCTIONISM? 5

CAT-LOGUE 1 11

CHAPTER 2: WHO IS READING THE BOOK OF LIFE? 15

CHAPTER 3: GENETICS: FROM GRAMMAR TO MEANING MAKING . . 19

1. The Complementarity of Syntax and Pragmatics 20

2. Is DNA a Language? 22

3. From Sequence to Function 25

4. Proteins: When Complexity Prevails 29

5. The Genetic System: Simple Mapping or a Complex Beehive? 34

6. The Reader's Role 38

7. Summary and Conclusions 39

CHAPTER 4: A POINT FOR THOUGHT: WHY ARE ORGANISMS IRREDUCIBLE? 41

Summary 41

1. Introduction. 41

2. Why Do We Need Signs? 42

3. Life's Irreducible Structure 43

4. Transcending the View "From Within" 45

5. Information as a "Difference That Makes a Difference" 48

6. Machines of Oblivion 49

7. Reductionism is Impossible Because We are Irreversible. 51

8. The Digital and the Analogical 51

9. Conclusions 52

CHAPTER 5: A POINT FOR THOUGHT: DOES THE GENETIC SYSTEM INCLUDE A META-LANGUAGE? 55

Summary 55

1. Introduction. 55

2. "Junk" DNA: Is It Really Junk? 56

3. Is Keeping the Junk "Energetically Favorable" to Deletion?. 58

4. ncRNAs as a Meta-Language 60

5. The Map and the Territory 61

6. Why Do We Need a Meta-Language? 64

7. Meta-Language is Inevitable 65

8. The Importance of In Between. 68

9. Conclusions 69

Cat-logue 2 71

Chapter 6: IMMUNOLOGY: FROM SOLDIERS TO HOUSEWIVES. 75

1. The Innate and the Adaptive 75

2. The Agents of the Specific Immune Response 79

3. Immune Recognition. 79

Chapter 7: A POINT FOR THOUGHT: IMMUNE SPECIFICITY AND BRANCUSI'S KISS 83

Summary 83

1. On Miraculous Drugs and Biological Specificity 83

2. Specificity in Immune Recognition 84

3. Specificity as Meaning Making 85
3.1. Recursive-Hierarchy 86
3.2. Synsymmetry 87
3.3. Hypothetical Inference. 90

4. Conclusions 91

Chapter 8: A POINT FOR THOUGHT: REFLECTIONS ON THE IMMUNE SELF 95

Summary 95

1. Introduction. 95

2. Danger! 98

3. A Reductionist Explanation of the Immune Self 99

4. Burnet and Saussure 105

5. Jerne and Peirce 108

6. Cohen and Volosinov 112

7. The Nose and the Finger 117

8. The Testes and the Immune Self 120

9. Conclusions 124

CAT-LOGUE 3 129

PART 2. WHAT IS THE MEANING OF THIS STORY?

CHAPTER 9: MEANING MAKING IN LANGUAGE AND BIOLOGY 133

1. Metaphorical Thinking 133

2. Living Systems and Boundary Conditions 135

3. Meaning Making 138

4. Meaning Making, Organization, and Disorganization 139

5. Meaning and Interaction 140

6. Implications 143

7. A Final Comment: "Let Truth Be Raised Up from the Ground!" 144

CHAPTER 10: GOD'S SACRED WORDS 147

1. The "Book of Life" and the Book of Genesis 155

CHAPTER 11: IT MEANS NOTHING 161

1. Life is Meaning Making 167

CHAPTER 12: A POINT FOR THOUGHT: MEANING—BRIDGING THE GAP BETWEEN PHYSICS AND SEMANTICS 171

Summary 171

1. Introduction 171

2. Information: Different Questions Lead to Different Answers 173

3. Information: A "Naturalistic" Perspective 174

4. Interactive Machines . . . 176

5. Maxwell's Demon . . . 177

6. Measurement as an Invention . . . 180

7. Maxwell's Demon in a Klein Bottle . . . 181

CHAPTER 13: THE REST IS SILENCE . . . 185

1. Wound Healing, Plasticity, and Context . . . 187

2. Epigenesis . . . 188

3. The Rest is Silence . . . 191

4. How Do Biological Systems Know Themselves? . . . 195

PART 3. ON THE WILD SIDE: FOUR LESSONS

CHAPTER 14: THE POLYSEMY OF THE SIGN: A QUANTUM LESSON . . 201

Summary . . . 201

1. Introduction . . . 201

2. Quantum Computing: The Liar and the Truth-Teller . . . 203
2.1 Quantum Machines . . . 204

3. From the Digital to the Analogue . . . 208
3.1 Codes in Natural Language . . . 209

4. Discussion . . . 212

CHAPTER 15: RECURSIVE-HIERARCHY: A LESSON FROM THE TARDIGRADE . . . 215

Summary . . . 215

1. Introduction . . . 215

2. Recursive-Hierarchy . . . 218

3. Organization Becomes Cause in Matter . . . 221

4. A Recursive-Hierarchical Metabolism? 222

5. A Lesson from the Physics of Computation 223

6. From the Baron von Münchausen to the Klein Bottle 225

CHAPTER 16: CONTEXT AND MEMORY: A LESSON FROM FUNES THE MEMORIOUS 229

Summary 229

1. Introduction: "Languaging" in Context 229

2. Immune Memory 232

3. A Lesson from Funes the Memorious. 235

4. Immune Memory and Aging 237

5. Conclusion. 238

CHAPTER 17: TRANSGRADIENCE: A LESSON FROM BAKHTIN 239

Summary 239

1. Introduction: There is No Alibi in Existence 239

2. Singularity in Language 241

3. Bakhtin on Meaning. 245

4. Bateson and the Mind. 246

5. The First Law of Human Perception 250

6. We are All Unique but Never Alone 254

7. From the "I" to the Other. 254

8. Signs as a Bridge between the "I" and the Other. 255

9. Conclusion. 256

CAT-LOGUE 4 257

PART 4. FROM MECHANICS TO POIESIS

CHAPTER 18: THE POETRY OF LIVING 261

CAT-LOGUE 5 271

References 273

Name Index 285

Subject Index 287

PART 1. HOW LOW CAN YOU GO? REDUCTIONISM AND ITS LIMITATIONS

Chapter 1 What is Reductionism?

Cat-logue 1

Chapter 2 Who is Reading the Book of Life?

Chapter 3 Genetics: From Grammar to Meaning Making

1 The Complementarity of Syntax and Pragmatics
2 Is DNA a Language?
3 From Sequence to Function
4 Proteins: When Complexity Prevails
5 The Genetic System: Simple Mapping or a Complex Beehive?
6 The Reader's Role
7 Summary and Conclusions

Chapter 4 A Point for Thought: Why are Organisms Irreducible?

Summary
1 Introduction
2 Why Do We Need Signs?
3 Life's Irreducible Structure
4 Transcending the View "From Within"
5 Information as a "Difference That Makes a Difference"
6 Machines of Oblivion
7 Reductionism is Impossible Because We are Irreversible
8 The Digital and the Analogical
9 Conclusions

Chapter 5 A Point for Thought: Does the Genetic System Include a Meta-Language?

Summary
1 Introduction
2 "Junk" DNA: Is It Really Junk?
3 Is Keeping the Junk "Energetically Favorable" to Deletion?
4 ncRNAs as a Meta-Language
5 The Map and the Territory
6 Why Do We Need a Meta-Language?
7 Meta-Language is Inevitable
8 The Importance of In Between
9 Conclusions

Cat-logue 2

Chapter 6 Immunology: From Soldiers to Housewives

1 The Innate and the Adaptive
2 The Agents of the Specific Immune Response
3 Immune Recognition

Chapter 7 A Point for Thought: Immune Specificity and Brancusi's Kiss

Summary
1 On Miraculous Drugs and Biological Specificity
2 Specificity in Immune Recognition
3 Specificity as Meaning Making
3.1. Recursive-Hierarchy
3.2. Synsymmetry
3.3. Hypothetical Inference
4 Conclusions

Chapter 8 A Point for Thought: Reflections on the Immune Self

Summary
1 Introduction
2 Danger!

3 A Reductionist Explanation of the Immune Self
4 Burnet and Saussure
5 Jerne and Peirce
6 Cohen and Volosinov
7 The Nose and the Finger
8 The Testes and the Immune Self
9 Conclusions

Cat-logue 3

Chapter 1

What is Reductionism?

Previously I raised some doubts with regard to reductionism as a strategy for understanding living systems. This short chapter aims to explain what reductionism is and why it is limited as a strategy for understanding living systems. Following this chapter, I briefly illustrate the limits of reductionism in genetics. This illustration will be followed by a more comprehensive introduction to genetics and by an in-depth discussion of the limits of reductionism in genetics.

As a general research strategy, classical reductionism assumes that a system under inquiry can be considered as hierarchical classification of objects in which the objects at each level are complex structures of the objects comprising the next lower level (Dupre, 1993). For example, any material object is composed of molecules that are composed of atoms. Reductionism involves the explanation of the objects at one level through the laws governing the lower level objects from which the previous objects are composed. For example, the structure of a crystal can be explained by the structure of the atoms that comprise it. Through the exchange of electrons the atoms create the macrostructure of a regularly ordered, repeating pattern extending in all three spatial dimensions.

Reductionism suggests that we should look "downward" and inquire the way in which the constituents of the system *determine* it *simply* in a bottom-up manner. I have no intentions of delving into philosophical casuistries concerning the meaning of *simply* and *determine*. Nevertheless, I would like to clarify that classical reductionism suggests that while trying to understand a given phenomenon we should break it into smaller components that are the cause in a relatively straightforward, direct and computable[1] manner, of the macro-level structure. For example, finding that matter is composed of conglomerates of atoms (*i.e.* molecules) is one of science's greatest reductionist achievements. A stone, an apple, a human being and a salt

---

[1] The relationship between computation and the impossibility of reductionism will be discussed in a later chapter.

crystal are all conglomerates of basic components that are combined according to the laws of physics. These components are drawn from a finite and well-defined set of atoms. To a certain extent the expression that we are all stardust is true but we are not only stardust. There is a difference between living systems and non-living systems and this difference makes a difference in the way we should approach these two different categories of nature.

As the etymology of reductionism teaches us, reducing is *bringing back*. Reductionism brings us back to the underlying and simpler level that determines the nature of the higher level of the system. There are different varieties of reductionism. Some of them are much more sophisticated (Dupre, 1993) than the classical and rather caricaturist version I presented but this variability does not change the basic definition of reductionism as previously presented.

In fact, setting aside that there are different varieties of reductionism, it is quite difficult to think about a scientific explanation that is *not reductionist*. Any scientific activity is reductionist in the sense that it aims to represent the complexity of a given phenomenon in a simpler manner. This is what explanation is all about and if the explanation/model is not simpler than the phenomenon it attempts to explain/represent then it is worth nothing. Therefore, the issue is not whether our scientific models and explanations should be simpler than the phenomena they try to explain but whether these simple models are of any help in *understanding* the system.

There is no doubt that there is an inherent problem in modeling a phenomenon. Dealing with the map (*i.e.* the scientific representation) rather than with the *territory* (*i.e.* reality) has its price. The price cannot be avoided or denied. Understanding is always a mediated process and modeling our world is an unavoidable step in understanding it. Without maps (cognitive or geographical) we are lost. The question is whether the map we hold is suitable for navigating us through the turbulent waters of the territory we are trying to understand. As I repeatedly argue, reductionism has limitations as a map for understanding the territory of life forms.

In *A System of Logic* John Stuart Mill (1911) already pointed to the elusive nature of explanation by describing it as "substituting one mystery for another." Sometimes our explanations are as elusive as what they try to explain. Instead of enacting a hidden layer of reality, as the etymology of explanation demands (*i.e.* ex (out) + planus (flat)), they substitute.

Even as a substitute an explanation has to pass the criterion of relevance. Sometimes explanations are an irrelevant substitute for a given "thought collective". For example, no matter how pluralistic we are, as modern scientists it will be impossible for us to accept the explanation that earthquakes result from evil spirits rather than from tectonic movements. This pre-scientific explanation is inadequate for the modern mind and therefore we cannot replace the mystery of earthquakes by the mystery of evil spirits.

For the scientist who is not a naïve realist it is impossible to avoid explanation as a substitution. We can never reach *the* ultimate explanation and our substitutions/explanations should be judged by their relative success in clarifying a variety of phenomena. However, substituting for the sake of substituting without using criteria for evaluating the quality of the explanation is a fallacy that should be avoided. According to Bergson, reductionism is an attempt to explain a system by substituting one mystery for another mystery that exists at a lower level of analysis. Bergson clarifies this argument saying that by pushing the problem a step backward, reductionism does not really explain the phenomenon although it arrogantly pretends to do so. This attack on reductionism seems to be over-stepping although it is definitely relevant for certain inferences drawn from a reductionist analysis. For example, explaining a human characteristic like aggression by identifying a gene of aggression is not really an explanation. A gene is defined as a "complete chromosomal segment responsible for making a functional product" (Snyder and Gerstein, 2003, p. 258). As such, the identification of this hypothetical aggression gene tells us nothing about the way and the path in which this chromosomal segment results in aggressive behavior. This gene may be used as a *differentiating characteristic* between more and less aggressive people but in no way it can serve as an *explanation* (or cause) of aggressive behavior.

Reductionism as a *general* scientific strategy is limited in many other senses, some of which have been intelligently presented by Dupre (1993) as well as many others. The main limitation of reductionism is that it cannot guide us in understanding the behavior of living wholes. It does not provide us with the appropriate strategy for understanding the behavior of functional living wholes. For example: Can the behavior, or the theories, of Richard Dawkins be explained by his selfish genes? I doubt it. Can the behavior of a human being be explained by understanding the laws governing the behavior of neurons in her brain? Again, I doubt it because it is clear to us that human behavior is social through and through, and that social processes exist at a higher level of analysis than neurons. We can play wise guys and argue that social processes are *represented* in the brain but this move leads us to nowhere. The whole world is represented in our mind! Theoretically, any human psychological or social trait may be finally traced to the brain or the genes. The question is how constructive this move is.

As summarized by Lewontin (1991):

> A lot of nature, as we shall see, cannot be broken up into independent parts to be studied in isolation, and it is pure *ideology* to suppose that it can. (p. 15; emphasis mine)

I emphasize the word *ideology* because science is intermingled with ideology more than the rhetoric of science would have us believe. In this context, reductionism is sometimes portrayed as the only legitimate scientific strategy and in many cases as the *only* scientific strategy. This arrogant position is theoretically ungrounded and its impoverishment is being masked by ideological claims. There is no doubt that modern science benefited in many ways by using a reductionist strategy, but these benefits cannot serve to justify the argument that reductionism is the one and the only possible approach to scientific understanding. Indeed, a knife is an extremely helpful tool for cutting vegetables, preparing tools from wood or slicing bread. However, inductively concluding that a knife is an excellent tool for picking your nose might be a dangerous conclusion. Overgeneralization is a typical reasoning fallacy. We should keep in mind that reductionism is just one tool in the intellectual toolkit of a scientist. The scientist should be careful that instead of using this tool she might find herself being used by the tool in the case that she begins to worship the tool. Human beings repeatedly make this error and turn tool using into the perversion of fetish. This perversion is evident in language use where people are ready to kill and to be killed for words. The general idea that words should serve people rather than that people should serve words ought to be extended to tools in general and to reductionism in particular. Again, this is not what happens in practice. In this context, the ancient Biblical prophets who passionately fought against the building of idols seem extremely relevant to our enlightened age in which tools of thought are being turned into idols.

Lewontin's statement suggests that it is ideology rather than common sense or rational thought that stands at the heart of reductionism. To illustrate the link between science and ideology, I would like to bring up an insightful observation by a great writer—Philip Roth. In one of his novels—*I Married a Communist* (Roth, 1999)—Roth comments on the hostility between politicians and novelists. Politicians, he says, like to speak in general and abstract terms: The *Nation*, *They*, the *Others*, *We* and *Justice* are just few key words from political speeches. In contrast, the novelist is interested in the more prosaic aspects of life in the sense that his novels are woven together from many small concrete details that interact in a complex way to constitute the plot, from the way a certain man gazes at a certain woman to the particularities of hand gestures in a certain context, and so on.

Scientists are like politicians in the sense that they transcend concrete reality in favor of overwhelming generalities. However, in contrast with politicians, scientists are judged by their ability to return from their abstractions to the prosaic, and not by their talent to sell these generalities or ideologies to the public. Newtonian mechanics would have meant nothing had it been unable to explain the free fall of an apple.

In this context, one should ask whether reductionism is the best strategy to bridge the gap between the prosaic aspects of living creatures and the abstractions of our science. There are instances in which the answer was originally "Yes!" but then turned out to be quite different. For example, the Genome Project sold both to the public and to stakeholders the illusion that sequencing the human genome would open for us the "book of life". In retrospective, one may be amazed by the religious rhetoric through which this reductionist venture was introduced to the public as the scientific move that would bring us to the new scientific dawn and to the full and complete understanding of living systems. Not only we are far from a complete understanding of living systems, but human welfare has not been dramatically changed as a result of this project which Lewontin (1991) describes as more "technical" rather than scientific. Life is still as elusive a concept as it was before the sequencing of the genes of various life forms was documented in their book of life.

Luckily, science as the collective, dynamic and creative activity of constructing knowledge and reflecting on knowledge does not always follow the theoretical abstractions of PR officers of biotechnology companies, armchair philosophers or ideologists of science. It simply moves on with clear indifference to the theoretical abstractions that attempt to explain its movement and to direct its progress. It is a creative activity and as such cannot be domesticated by scholarly reflections.

Philosophical abstractions, reductionism among them, appear *post hoc* and might mislead the scientific activity by providing it with an inappropriate representation of its own activity. Psychology may teach us an important lesson about the way once distorted image might be harmful, and reductionism is just a concrete example.

What is interesting to mention again is that the deficiency of reductionism as the ultimate *strategy* has been masked through the success of reductionism as a *local tactic* in certain fields of scientific activity. Indeed, there are enormous benefits in breaking a system into its constituting components. In fact, it is even quite trivial to realize that by adopting a reductionist approach we learn a lot about the components of a system and about local interactions between the components. However, those local battles do not indicate the win of the war. As we all know one may win the battle but lose the war. *Breaking a living system into its components is not enough*. After breaking the system one has to understand how it works as a whole. Otherwise she will be left with a technical and local understanding which has no validity for understanding interactive processes that constitute the living system as a functioning whole. This conclusion is clear to those who encounter living systems with all their complexity *in vivo* and it will be clear to the reader too as the book unfolds.

Cat-logue 1

The discussants: A university professor who is the author of the current manuscript (Dr. N) and his white cat, Bamba.

Dr. N: Hello Bamba! What are you doing?

Bamba: Just reading your short introduction to reductionism.

Dr. N: And what do you say? Isn't it a clear and concise introduction to reductionism?

Bamba: Well ... do you know that *ignorance* has etymological roots in *ignotus* meaning unknown?

Dr. N: I am impressed by your knowledge but what are you trying to say? I have the feeling that you are going in circles instead of giving me a straight answer.

Bamba: This is an interesting metaphor. Circular! Straight! As a hunter in nature I always try to avoid straight lines even if they are lines of thought. Straight lines are good only when you jump on your prey or when you see the answer in front of your eyes. When you are inquiring into an issue the topology of your voyage does not allow you the straight lines of Euclidean geometry. You see, hunting ideas is a serious issue, no less than hunting mice.

Dr. N: Are you ready to explain my ignorance and its relevance for the discussion on reductionism?

Bamba: I was wondering whether you read Fleck's *Genesis and Development of a Scientific Fact* (Fleck, 1979).

Dr. N: Fleck? Who is Fleck?

Bamba: Ludwik Fleck was a Jewish-Polish physician and biologist who wrote in 1935 a book that is the cornerstone for the social study of science. This revolutionary work predated Kuhn's famous book and Thomas Kuhn even wrote the foreword to the English edition. Fleck may add a depth to your scholarly discussion of reductionism.

Dr. N: Hmm ... Sounds interesting, in what sense?

Bamba: You see Fleck argues that an explanation can survive and develop within a society only if it is "stylized in conformity with the prevailing thought style" of the society.

Dr. N: What does he mean by a "thought style"?

Bamba: To understand this concept you should be familiar with another key concept in Fleck's book—"thought collective." A thought collective is a community of people mutually exchanging ideas and maintaining intellectual interaction. A thought style is the given stock of knowledge used by the thought collective. For example, the community of psychoanalysts maintains a thought collective, which is characterized by a certain thought style. They refer to specific texts, use a certain set of keywords, and share the same knowledge of what is personality.

Dr. N: Ok. So what is the thought collective that maintains reductionism and what is the thought style that characterizes this collective?

Bamba: It is difficult to define this thought collective but Dawkin's *The Selfish Gene* is clearly a sacred text for this collective and neo-Darwinism is a part of their thought style.

Dr. N: Does Fleck have insights relevant for understanding reductionism? It is not quite clear how a simple tactic of research such as reductionism turned into a strategy, or worse, a sacred ideology.

Bamba: Fleck may explain this process by saying that ideas acquire "magical power" and "exert a mental influence simply by being used." In themselves ideas are not problematic. One can break a system into components in order to understand it. One can use intuition, imagination, and whatever he or she wants. However, ideas are constituted and legitimized in the social arena. This is the place where they turn into ideological "isms" that acquire magical power. This is the place where the dynamic idea turns into a dogma, an authoritative and obligatory code of belief. Even worse, when the elite of a thought collective encounters the masses, the *Volgos*, it communicates with them by adopting a kind of a simplified rhetoric that ignores the complexities and the qualifications known to the in-group circle. Fleck said that in the case where the elite does not enjoy a strong position in society the stronger their bond with the masses will be, and the more simplified their rhetoric will be. This "vulgarization of science" may turn into a threat when newcomers to science or even practicing scientists themselves start to accept the rhetoric originally created for the mass as their own scientific dogma.

Dr. N: Now I understand how a popular science writer like Richard Dawkins turned into the ultimate representative of scientific reductionism. In a post-modern culture where the great narratives have collapsed and the intellectual elite is under continuous attack, vulgar science prevails.

Bamba: What a revelation. I told you to read Fleck.

Dr. N: How do you explain the fact that he is almost unknown?

Bamba: It's even worse. Not only is that he is almost unknown, even Thomas Kuhn, who admitted an intellectual debt to Fleck, underestimated him. Kuhn wrote the introduction to Fleck's book in the 1970s, at the time of the cold war. My hypothesis is that the idea of the thought collective had the flavor of the hatred of Marxism that indeed underlies Fleck's treatise. Moreover, the idea of the thought collective is in sharp contrast with the American ethos of individualism, which is of course another "ism" of a thought collective. Fleck reminds us, to use a Bakhtinian expression, that *we are all unique but never alone.*

Dr. N: Hmm ... an interesting explanation. By the way, and from a reflective stance, what is your thought style?

Bamba: I would love to discuss this question after dinner. My selfish genes urge me to supply them with their daily amount of energy, and I notice that we have salmon for dinner, my genes' favorite food.

Dr. N: Are you a reductionist?

Bamba: Not at all. But I am afraid that my genes are.

Chapter 2

Who is Reading the Book of Life?

Genetics provides us with a good case for illustrating the poverty of reductionism and for understanding the importance of understanding living systems as meaning-making systems. In this context, the Genome Project is a wonderful case for achieving these aims because at the beginning it was brimming with reductionist enthusiasm that easily slipped into the rhetoric of brute determinism. We were told that in due course the book of life would be available for reading and that the destiny of each of us could be found in this book. An illustrating statement is the one of DeLisi:

> This collection of chromosomes in the fertilized egg constitutes the *complete set of instructions* for the development, determining the details of the formation of the heart, the central nervous system, the immune system, and every other organ and tissue required for life. (DeLisi, 1988, quoted in Nijhout, 1990, p. 441; emphasis mine)

In retrospect this reductionist and deterministic enthusiasm seems as if it was taken from one of those fiction stories written by the great Argentinean writer Jorge Louis Borges; an imaginary story about a kingdom in which the destiny of each subject was predetermined by his "genetic book of life". I can imagine Borges opening this fiction story by writing something like this:

> Arodonos of Aropega (b. 585) describes in his book *Tractatus Genomica* the lost kingdom of Genoma in which each person was born with his book of life. In fact, this kingdom was no more than a living library in which the features, characteristics and destiny of each person was determined by the holy letters, written by the blind author known to his disciples as Snikwad.

Irony was always one of my favorite rhetorical tactics and I hope that Prof. Dawkins/Snikwad will not be hurt. This irony however is clearly grounded in the ideology represented in reductionist texts like *The Selfish Gene*. This ideology is clearly presented and criticized in Lewontin's *Biology as Ideology* (1991) and Keller's *The Century of the Gene* (2000). Not so surprisingly, the reductionist ideology encountered the complexity of the world and:

> Contrary to all expectations, instead of lending support to the familiar notions of genetic determinism that have acquired so powerful grip on the popular imagination these successes pose critical challenges to such notions. (Keller, 2000, p. 5)

The idea that understanding the DNA sequence will allow us to read the book of life turned out to be an illusion, a kind of logical fallacy that equates the working whole with its components: *Pars pro toto*. It did not take a long time to realize that:

> The causal pathway [from genes to organism] is endless and involves not only genetic but manifold structural chemical and physicochemical event, a defect in any of which can derail the normal process. (Nijhout, 1990, p. 442)

The conclusion at least as summarized by Keller points at *meaning* as a key concept for the post-genomic era:

> But now in the call for a functional genomics, we can read at least a tacit acknowledgement of how large the gap between genetic "information" and *biological meaning* really is. (Keller, 2000, p. 8; emphasis mine)

The reason why the term "information" appears in quotation marks is not trivial. Information is another metaphor in the genetic research but a problematic metaphor. As argued by Nijhout (1990),

> to apply information theory in a proper and useful way it is necessary to identify the manner in which information is to be measured (the units in which it is to be expressed in both sender and receiver and the total amount of information in the system and in a message), and it is necessary to identify the sender, the receiver and the information channel (or means by which information is transmitted). As it is, there exists no generally accepted method for measuring the amount of information in a biological system, nor even agreement of what the units of information are (atoms, molecules, cells?) and how to encode information about their number, their diversity, and their arrangement in space and time. (p. 443)

Biological information is a concept that I will discuss later, but let us ponder more on Keller's statement. The above quotation from Keller is a highly important statement that points to the gap between meaning and information. Meaning cannot be simply extracted from information (whatever it is) or the mere sequence of letters. Information can be the infrastructure for meaning, it can constrain the meaning we may attribute to a message, but meaning cannot simply pop-up from a sequence, whether a genetic sequence or a sequence of words. Something is missing and this "missing link" is crucial for our understanding of living systems. Unfortunately, I recurrently encounter computer scientists who still believe with a religious zealousness that meaning can be reduced to information and that it is only a matter of time until scientific advancement will prove their point. Those "scientists" should read what Lewontin wrote years ago:

> A deep reason for the difficulty in devising causal information from DNA messages is that the same "words" have different *meanings* in different *contexts* and multiple functions in a given context, as in any *complex language*. (Lewontin, 1991, p. 166; emphasis mine)

This is an important argument that may help us to understand why meaning cannot be simply extracted from a sequence. As will be later explained meaning is the outcome of a contextual event and not an encrypted message. The next chapter concisely introduces basic ideas of genetics and delves more deeply into the shortcomings of genetic reductionism in encountering the complexity of living systems.

Chapter 3

Genetics: From Grammar to Meaning Making

DNA is a long chain, which is constructed out of basic units. These subunits are nucleotides—molecules made up of a sugar (deoxyribonucleic) attached to a single phosphate group and to a base which is A (Adenine), G (Guanine), C (Cytosine), or T (Thymine) that give us the four "letters" of DNA. In Fig. 3.1 you can see a schematic representation of the basic unit.

Amazingly simple! And to think that life forms are grounded in just few genetic letters! Well the idea that the world was created by a finite set of letters is not so new after all and, surprisingly as it may sound, one can easily find it in one of Judaism's mystical texts.

The alphabet of creation is one of the legends told in *Sefer Ha-Zohar* (*i.e.* The Book of Splendor). This ancient mystical text, which was originally written in Aramaic, dates back to the thirteenth century when a Jewish Spanish scholar wrote it based on knowledge that was proclaimed to an ancient Jewish Rabbi. This legend is very nice because it shows that the idea

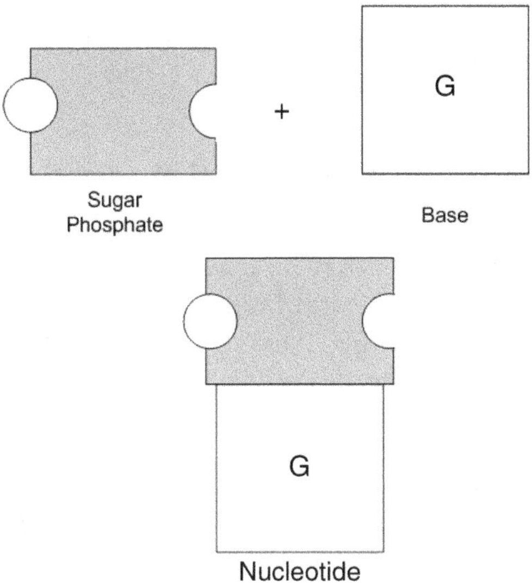

Fig. 3.1 A schematic description of the nucleotide.

of creation through a finite set of letters/tokens is in itself not the original discovery of genetics research. A summary of the legend follows:

> Twenty-six generations before the creation of the world, the twenty-two letters of the [Hebrew] alphabet descended from the crown of God whereon they were engraved with a pen of flaming fire. They gathered around about God and one after another spoke and entreated, each one, that the world be created through him. (Shahn, 1954, p. 1)

Sefer Ha-Zohar suggests that the world was created through the Hebrew alphabet, which is described not as a set of arbitrary symbols but as a group of tokens immersed with life. The homomorphic description of the Hebrew alphabet should not distract us from our main argument and as will be shown later it definitely converges with this book's main argument.

It seems that any act of construction we are familiar with, from sentences in natural language to the structure of proteins and organisms, involves a finite set of units/letters that serves as the base for the construction process. This idea is simple but far from being trivial. How is it possible to understand the complex and contextual nature of real living systems with the "mathematical" and decontextual sequence of a finite genetic alphabet? Can we reduce the mystery of life to the genetic alphabet? Are we moving in a reductionist direction?

To address these questions it is worth examining the relation between the abstract "mathematical" nature of the genetic language and its contextual and pragmatic counterparts. This examination will be done through insights gained in linguistics.

1. The Complementarity of Syntax and Pragmatics

The linguist Roman Jakobson once commented that with regard to the relation between structures that are *context-independent* and *context-dependent* mathematics and natural language are bi-polar systems. Let me explain this argument. There are structures that are context-dependent and there are structures that are context-independent. Mathematics and natural language are the prototypical fields in which we can find these two types of structures.

Mathematics involves decontextualized structures since it is based on a *syntactic* form of representation. A syntactic form of representation describes the *rules* that govern the behavior of a set of letters (or tokens), abstracting both from their semantics (their meaning) and their pragmatics (the way they are actually being used).

A syntactic form of representation has several properties that differentiate it from other forms of representation. The defining property of the syntactic form of representation is that it is characterized by *meaningless* tokens/letters of an alphabet. We may define a meaningless token as follows:

A token is meaningless when its contribution to the whole of which it is a part is totally determined by the rule(s) (the grammar) that connect it with other tokens that compose the whole, and not by its pragmatics, which is always particular and unique.

For example, consider the familiar logical syllogism:

1. IF **A** THEN **B**
2. **A**
3. THEREFORE **B**

Whatever is the meaning of the tokens **A** and **B**, and no matter who produces this syllogism or who is the addressee of this syllogism, the structure of the syllogism is always valid. Therefore, whether **A** means "the cat is white" or "the dragon is pink" is of no importance.

Natural language includes a syntactic context-independent aspect—its grammar. However as it is practically used by human beings it is *context-dependent*. When using language for communicative means, we heavily rely on contextual cues. For example: Who is producing the utterance? Who is the addressee of the utterance? When is it produced? What is the common knowledge of the communicating agents? and so forth. In natural language, as it is used *in vivo*, pragmatics rather than syntax is the salient feature.

Here we get into the point. Jakobson insightfully commented that what is interesting is that the two bi-polar systems, the mathematical and the linguistic, the syntactic and the pragmatic, are *complementary* in the sense that each of them is the most appropriate *meta-language* for the structural analysis of its companion: Mathematics as meta-language for natural language, as illustrated, for example, by the work of Zelig Harris (1968), and natural language as a meta-language for mathematics.

Jakobson's insight should be explained with regard to the necessary interdependence between the abstract and decontextual aspect of language, what Saussure described as *la Langue*, and the contextual pragmatic event of *languaging* (Maturana and Varela, 1992), what the famous linguist Ferdinand de Saussure described as *Parole*. It seems that in practice living systems are based on the complementary aspects of two modes of operation: a syntactic, mathematical and decontextual mode and a pragmatic and contextual mode. Trying to isolate one of the modes and to see it as the whole picture is a wrong move. Whether in linguistics or biology, *downward reductionism* to grammar or *upward reductionism* to metaphysical concepts is a wrong move. Later we will delve into this issue more deeply by inquiring into the relationship between language and meta-language in the genetic system but in the meantime let us return to the structure of DNA.

2. Is DNA a Language?

DNA is composed of two anti-parallel strands. Each strand is composed out of covalently linked through phosphodiester bonds connecting the phosphate in one nucleotide to the sugar in the next nucleotide. So what we actually have in each strand is a chain of sugar–phosphate–sugar–phosphate and so on (Fig. 3.2).

The two strands are held together by hydrogen bonds between complementary bases in each strand. A hydrogen bond refers to a state in which the positively charged region of one water molecule comes close to the negatively charged region of another water molecule. The result of this encounter is a weak bond (non-covalent bond) known as a hydrogen bond (Fig. 3.3).

To review, a water molecule is composed of two atoms of hydrogen and one atom of oxygen. The oxygen atom is strongly attractive to electrons (the negatively charged particle of the atom), and the hydrogen atom is only weakly attractive to the electrons. As a result the electrons have an unequal distribution in the water molecule, where the hydrogen atoms are positively charged and the oxygen atom is negatively charged. Water molecules interact with each other and can attach to each other. This is the reason why we can surf on a wave. Give up hydrogen bonds and beach boys in California would have to improve their surfing skills and surf on isolated molecules.

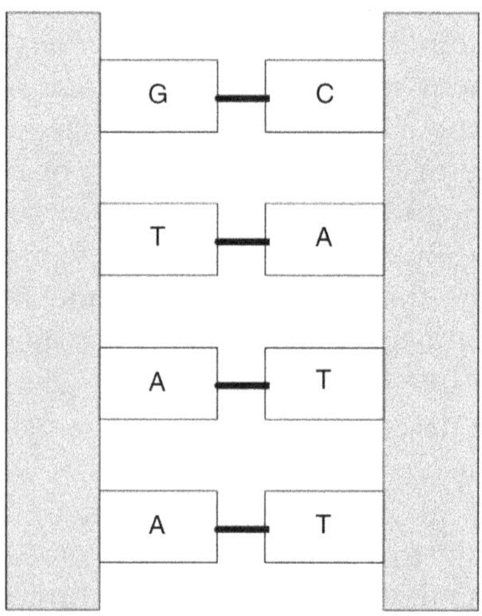

Fig. 3.2 A schematic representation of the DNA.

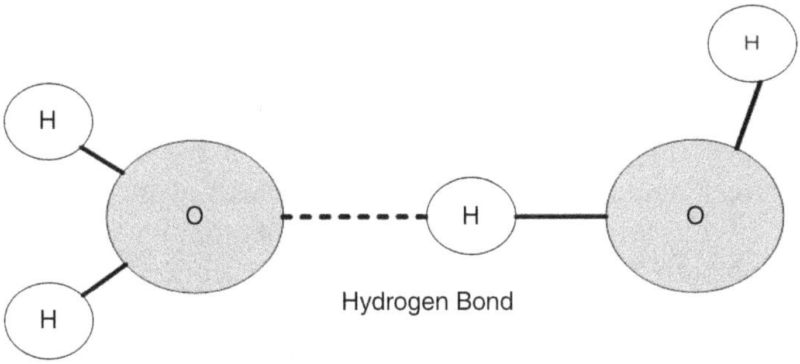

Fig. 3.3 A schematic representation of the hydrogen bond.

It is important to realize the role of non-covalent (weak) forces in nature. Non-covalent forces bind atoms without the exchange of electrons. Non-covalent forces allow flexibility and the possibility of separation. At least in the biological realm the Catholic statement: "*Until death do us part* ..." luckily does not hold. Without the weak forces holding the two strands together separation and therefore heredity would have been impossible. Thank God that the logic of the DNA is not the logic of the Catholic Church. I am mentioning this issue of weak forces because, as will be later discussed, weak forces are crucial for a variety of meaning-making processes both in natural language and in biology.

Back to the concept of DNA: We have two strands attached to each other and those strands wind around each other to create the famous double helix. The bases are not randomly attached. A always pairs with T, and G with C. As I previously mentioned, the four bases are usually considered to be the "letters" of the genetic alphabet. The linguistic metaphor is evident in this case and for the time being I do not want to discuss the question of whether the use of "letters" to describe the four bases is a use that points to a deep similarity between natural language and genetics. For a better understanding of the defining characteristics of language and the analogy between linguistic and genetic letters we should now turn to Harris.

In *Mathematical Structures of Language*, Harris (1968) identified what he considers "universal and essential properties of language" (p. 6). The first property is that the elements of language are discrete and arbitrary in the sense that the sounds out of which words are composed do not suggest the meaning of a word. Later we will discuss this property as the defining characteristic of the *digital code*. Another property of language is that *not all combinations* of the discrete units occur. This property allows us to define larger constructions as restrictions (or constraints) on the combination of lower-level tokens (*e.g.* words). These properties are clearly evident in the genetic system: The elements are discrete, linearly ordered, and arbitrary in the sense that the chemicals that comprise them do not indicate their function. However, if we ignore the pragmatics of language then there are problems in applying the linguistic metaphor to the genetic system.

The reader should keep in mind that for each language we must assume an *observer* or *interpreter*—an actual or hypothetical being that may make sense out of the letters' sequence. This interpreter does not have to be a human being but as long as we stick to the linguistic metaphor we must assume the existence of someone or something that can use a sequence by turning it into a "difference that makes a difference". In other words, someone or something must translate a *sequence* into *a* function, information into meaning, or syntax into pragmatics. This is a challenge we shall address later in the book. For now, let us stick to the popular and oversimplified conception that the genetic alphabet is a part of the language that underlies heredity and provides the "instructions" for the construction of proteins.

Proteins are amazing molecules that are the building blocks of the living systems. Later, I will devote a whole section for discussing about them. In the meantime a concise description will suffice. Proteins are molecules that are composed of amino acids. Amino acids have a carboxylic acid group and an amino group linked to a single carbon atom called the α–carbon. Proteins come in an impressive variety that results from the side chain attached to the α–carbon. They are unordinary molecules and as one learns more and more about them one understands why they underlie so many biological structures and processes. Figure 3.4 is a schematic description of a protein.

The most basic structure of a protein is determined by the sequence of the amino acids from which it is composed. Here we start to sense the relation between the genetic sequence and the sequence of proteins. If we have a *linear sequence* composed of four basic letters and a linear sequence composed of twenty letters (each signifying an amino acid) then we should try to understand the way *one sequence is mapped onto the other sequence*. The linguistic metaphor inevitably pops up again. Isn't it a classical problem of translation like the one of translating from one language to another language? I advise the reader to be both excited by and critical of the linguistic metaphor in biology. Later I will explain why. At this point, I would like to turn again to DNA strands and to heredity.

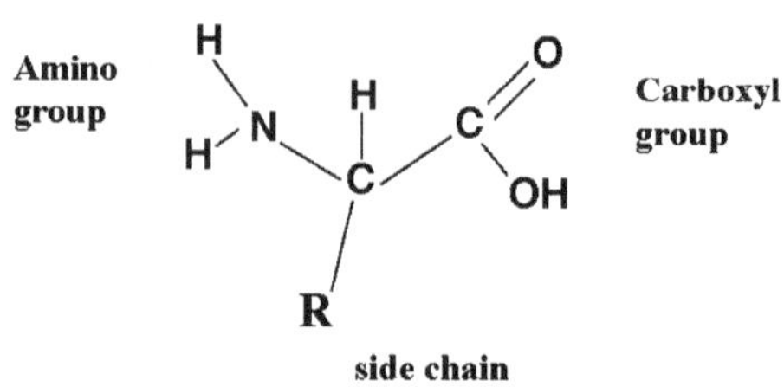

Fig. 3.4 A schematic representation of a protein.

3. From Sequence to Function

Turning a sequence of bases into a protein is a highly complex process, which is orchestrated by a dynamic network of molecular agents. In this context, the downward reductionism that excited the public's mind at the beginning of the Genome Project does not seem to have firm basis in our current biological knowledge. But why? After all we do know that in a causal chain, albeit highly complex, the DNA sequence does lead to structures that in their turn constitute the organism. In this context, why is it the case that organization, structure, or function cannot be simply reduced to the sequence? Is it just our current state of knowledge that prevents us from completing this reductionist venture or is it impossible *in principle*? It is rather difficult to find a criterion that differentiates between the kind of impossibility that results from our current state of knowledge and the impossibility that results from the state of the world as it is (*i.e.* impossibility in principle). Later, in Chapter 4 entitled "Why Are Organisms Irreducible?" I will provide my original thesis why reductionism is impossible in principle. Meanwhile we should acknowledge again that *meaning* is a key term in answering this question.

To better explain the poverty of reductionism, specifically with regard to the transformation from DNA to proteins, we should be familiar with the idea that reality is *layered* in a hierarchical structure in which higher levels are complex conglomerates of lower levels. The idea that the world of phenomena may be sorted into *qualitatively* different categories that reflect levels of complexity may be of great value in answering our question whether it is impossible to reduce structure to sequence or meaning to information.

The philosopher C. S. Peirce introduced the idea of a layered reality by using his three categories of Firstness, Secondness, and Thirdness. He introduced the concept of *phaneron* in order to explain these categories. By phaneron Peirce simply means

> the collective total of all that is in any way or in any sense present to the mind, quite regardless of whether it corresponds to any real thing or not. (CP 1:284)[1]

In other words:

> I use the word *phaneron* to mean all that is present to the mind in any sense or in any way whatsoever, regardless of whether it

[1] CP(x:xxx) refers to CP (volume:paragraph).

> be fact or figment. I examine the phaneron and I endeavor to sort out its elements according to the *complexity of their structure*. I thus reach my three categories. (CP 8:213, c.1905; emphasis mine)

This idea means that our reality whether the physical, the psychological, or the mathematical is layered in an increasing order of complexity and that this complexity should be discussed using *qualitatively* different categories. For example, in mathematics we have sets that are collectives of objects. The set of fruits includes members such as apple, mango, or grapes. Sets can be members of higher-order categories—Classes. *Classes* are entities that have sets as their members, and *conglomerates* is a category that was created to deal with collections of classes. Therefore, we have a hierarchy of categories that are qualitatively distinguished from each other. By qualitatively distinguished I mean that the properties that characterize the higher levels of the hierarchy cannot be trivially and analytically deduced from the properties that characterize the lower levels. This statement may be interpreted in several legitimate senses, for example, that the description of properties that exist on one level cannot be exhausted by the concepts of a lower level. Let me illustrate this point. The cell is composed of molecules that are composed of atoms that are composed of smaller particles known as quarks. However, in order to describe the behavior of the cell we must introduce new concepts and terminology that cannot be reduced to the terminology of particle physics.

Bertrand Russell (1908) was one of those who directed our attention to the idea that reality is hierarchically layered. Therefore, I will dedicate some space to present the basic tenets of his Theory of Types. Russell's point of departure is certain *contradictions* or paradoxes such as the liar paradox:

> Epimenides the Cretan said that all Cretans were liars, and all other statements made by Cretans were certainly lies. Was this a lie?

If the answer to the above question is positive and the Cretan is lying then he is speaking the truth and he is a liar, which means that he is speaking the truth and so on *ad infinitum*.

As noted by Deleuze (1990) the meaning and etymology of *sense* is closely associated with the notion of pointing in a certain direction. Paradox is senseless since it oscillates infinitely between two diametrically opposed values (true and false) without being able to peacefully rest on one of them. It does not point in a single and clear direction.

Another paradox that I find mind turbulent is Russell's Paradox:

> Let *w* be the class of all those classes which are not members of themselves. Is the class *w* of those classes which are not members of themselves a member of itself?

Most classes are not members of themselves. The class of horses is not a horse. The class of nuns is not a nun and so on. Let us return to the above question. If the answer to this question is positive and *w* is a member of itself then *w*, which is defined as the class of classes that are not members of themselves, cannot be a member of itself. If the answer is negative and *w* is not a member of itself then it is without any doubt a class which is not a member of itself and therefore clearly a member of itself…!

Try to think about this amazing paradox after drinking a cup of Irish coffee and you will sense its turbulent nature. Why using narcotics when a good logical paradox and a cup of Irish coffee may create a similar synergetic effect!

Russell's paradox gave mathematicians a headache and threatened to shake their secure world. Russell considered the paradox as illustrating a *reflexive* fallacy (Russell, 1908, p. 230) in which the totality is considered in terms of its components. He tried to solve the paradox by introducing type theory (Russell, 1903). According to type theory, the source of the paradox is the assumption that classes and their members form a single, homogenous logical type. In contrast, he proposed that the universe should be classified into a *hierarchy of types* (*e.g.* individuals, classes of individuals, and so on), by defining a type as "the range of significance of a propositional function, *i.e.* as the collection of arguments for which the said function has values" (Russell, 1908, p. 236). In this context, members of a class must be drawn from a single logical type. Russell's paradox cannot arise in a context where members must be of the same logical type.

Russell elaborated on type theory in order to deal with semantic paradoxes. His ramified theory suggests that the hierarchy of types is supported by a hierarchy of properties, which have a range of signification. In this context, Russell introduced the *vicious circle principle* (VCP), which censures self-reference: "no totality can contain members defined in terms of itself" (Russell, 1908, p. 237).

Although, the type theory in its different versions supports us with a possible approach for studying multi-level systems, it proscribes self-reference which is an essential characteristic of living systems. After all, it was argued that the most significant property that defines living systems is that they are *autopoietic* (Maturana and Varela, 1992), being able to produce themselves.

Russell's ideas should be critically examined from another perspective. As argued by Ben-Jacob (1998):

> Our logic and mathematics are based on the notion of a set composed of elements. Implicitly, the set is closed and static, the elements have fixed identity (it does not change due to the fact that they are part of the set) and they either do not have an internal structure or, if they do, it is not relevant to the definition of the set. (p. 67)

This logic of sets that leads to paradoxes is irrelevant to the study of living systems that are composed of totally different kinds of elements. It was argued in Ben-Jacob (1998) that in the context of organisms, alternatives do exist to Russell's set-oriented perspective on layered systems.

As will be argued later, it seems that paradoxes are necessary to the existence of semiotic systems in general and natural language in particular. For the time being I do not want to shift the discussion and we may use Russell's idea concerning the layered nature of reality while at the same time acknowledging that reality has a recursive nature.

Having the idea of a layered reality in our mind we may turn back to the issue of sequence and structure. A sequence is a collection of objects (*e.g.* differentiated letters) that has been ordered such that each member either comes before or after every other member. It is not a simple collection of elements. That is, a sequence assumes (a) an *alphabet* and (b) a relation of precedence between the letters comprising the alphabet. That is we have *differences* (*i.e.* different tokens), and *repetitions* of these tokens along a single dimension. Differences and repetitions are two highly important terms that will be used in the concluding part of the book and one should keep them in mind. They are also the constituting features of a Turing machine (TM), which epitomizes the idea of a computing machine. To review, a TM is composed of a tape that is divided into cells. Each cell contains a symbol from a given finite alphabet that includes the blank symbol. Although the alphabet is finite the tape is not and it is potentially extendible to the left and to the right. DNA is a sequence since it is composed from a basic alphabet of four letters and a blank symbol (is the blank symbol an intron?). Is the DNA sequence the tape of the biological TM?

Turing had a significant influence on formal language theory. In formal language theory languages are nothing more than "sets of strings drawn from some alphabet" (Searls, 2002, p. 212). This idea has also influenced the linguistic metaphor in biology. After all, if the DNA is no more than a string "drawn from some alphabet" then it is legitimate and scientifically reasonable to speak about the "language of the genes" (Searls, 2002).

The difficulty with this formal conception of language is that it is not clear how to move from the sequence (*i.e.* the string drawn from alphabet or the tape of the TM) to structure and function (*i.e.* meaning). In the case of genetics, the question is how to move from a sequence of DNA (*i.e.* a segment of the DNA) to the structure of a given protein, from the language of nucleotides to the language of amino acids, and at a higher level of analysis to the biological interactions that underlie the living organism. To answer this question we should understand what makes a sequence a distinguished type or category from structure.

We may argue that a structure is irreducible to sequence because to translate the sequence into a structure the system must transcend its boundaries and move from one level of organization to a qualitatively different level of organization. At face value this move should not be a problem. After all biological systems do it all the time. This is true but it means that in order to understand this move a reductionist approach would not suffice. To better understand the layered nature of biological systems and the difficulties associated with the modeling of moves between different levels of biological organization we turn to the structure of the protein and its relation to the genetic sequence.

4. Proteins: When Complexity Prevails

In the public's mind proteins are associated with the materials body builders consume in order to increase their muscles mass. However, the use of proteins by organisms is much wider than the public may imagine and proteins are indispensable molecules for the living organism. For example, enzymes are proteins that catalyze reactions. There are transport proteins such as hemoglobin, which carries oxygen. Signaling proteins like hormones carry messages, and gene regulatory proteins are involved in genes expression. There are approximately 100,000 different proteins in the human body, each protein with its unique three-dimensional arrangement.

To review, a protein is actually a molecule, which is composed from a sequence of amino acids. A given type of protein is composed from the same number of amino acids in the same order and proportion. For example, insulin is composed from 30 glycine + 44 alanine + 5 tyrosine + 14 glutamine + ⋯ In other words it is a biological structure which is basically composed from a sequence. The sequence of the insulin begins as follows: MALWMRLLPLL… These linear chains of polypeptide fold to generate the three-dimensional structure of the protein.

The function of a protein is dependent on its structure since it operates by binding to another molecule. Therefore it may be of interest to determine the structure/function of the protein by using the linear sequence. Is it possible?

Currently the answer is negative and I will explain why. In this section, I would like to explain what proteins are and illustrate the limits of reductionism through the problem of predicting the structure/function of a protein from its sequence.

All amino acids have in common a central carbon atom (C_α), which is attached to a Hydrogen atom, to an amino group, to a carboxyl group, and to one of 20 types of side chains (Fig. 3.5).

During the synthesis of a protein these amino acids join together by forming *peptide bonds*. These bonds occur when one carboxyl group joins the amino group of another amino acid by eliminating water (Fig. 3.6).

As we can see, the chain is created when a peptide bond is established between the C′ of one residue and the nitrogen atom of the next. This chain is the backbone of the protein. The backbone is quite rigid due to the

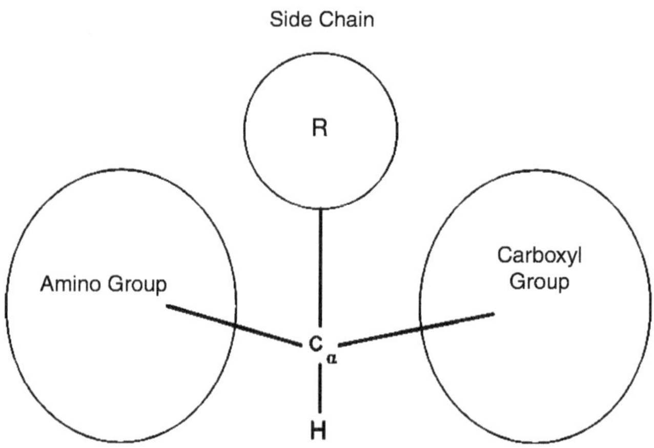

Fig. 3.5 A schematic representation of an amino acid.

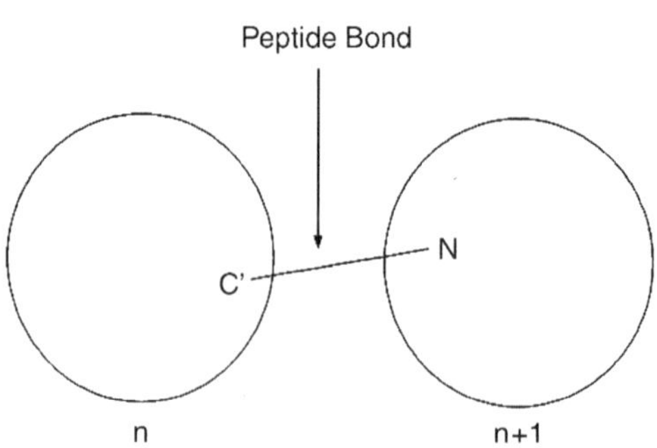

Fig. 3.6 A schematic representation of a peptide bond.

covalent forces at the C_α atoms. However, this relative rigidness does not mean that the backbone is static. The peptide units can *rotate* around the C_α–C' bond (with an angle of rotation symbolized by the Greek letter phi ϕ) and the N–C_α bond (with an angle of rotation symbolized by the Greek letter psi ψ). The ability of the units to rotate is important to understand since the protein turns out to be a dynamic organization rather than a static structure even at its primary level. Indeed it was found that the protein exhibits a range of motions ranging from local atomic fluctuations to complex global rearrangements. We will turn to this point later but even at this stage the reader should notice that the protein exhibits a delicate and well-orchestrated balance between order and disorder, static and dynamic.

So far things look rather simple. After all if we have a sequence, what is so difficult about predicting the structure? The protein has several *levels of organization* that make predicting its final structure an extremely complex problem. The level we discussed so far is the primary level that describes the protein as a sequence of amino acids. However, there are three additional levels: Secondary, Tertiary and Quaternary. Figure 3.7 shows a schematic description of a tertiary structure.

Those levels turn the protein into a highly complex and dynamic organization quite different from the simple and symmetric structure of the double helix. Kendrew, who was the first to determine myoglobin structure in 1958, expressed his surprise at the complexity of the protein by saying:

> Perhaps the most remarkable features of the molecule are its complexity and its lack of symmetry. The arrangement seems almost totally lacking in the kind of regularities which one instructively

Fig. 3.7 The complexity of the tertiary structure.

> anticipates, and it is more complicated than has been predicted by any theory of protein structure. (Quoted in Branden and Tooze, 1999, p. 13)

Please remember Kendrew's surprise at the protein's lack of symmetry since it is one of the issues I plan to discuss in a chapter dealing with "The Specificity Enigma" in immunology. Proteins do exhibit regularities as is evident, for example, in its secondary structure. The secondary structure of the protein comes either as *alpha helices* or as *beta sheets*. See Figs. 3.8 and 3.9 for schematic representation.

These strands result from the molecules needed to pack hydrophobic side chains into the interior of the protein and they are realized through hydrogen bonds between the main chain NH and C′=O.

The secondary structures are organized in *motifs* by packing side chains from neighboring α helices or β strands close to each other. These motifs combine and form globular structures known as *domains* and these domain structures form the tertiary structure of the protein. There are many proteins that are organized at a higher level in which several identical polypeptide

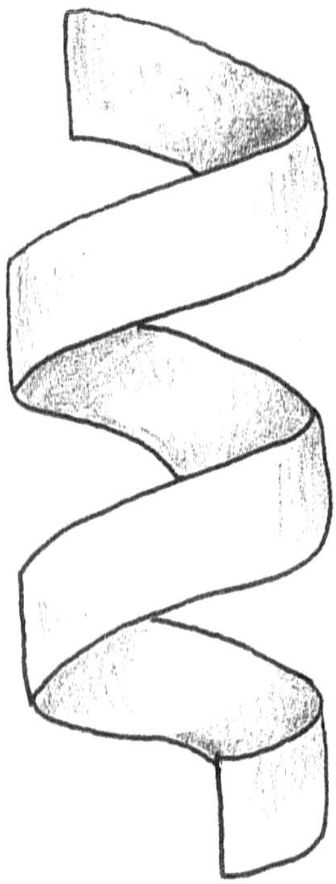

Fig. 3.8 A schematic representation of the alpha helix.

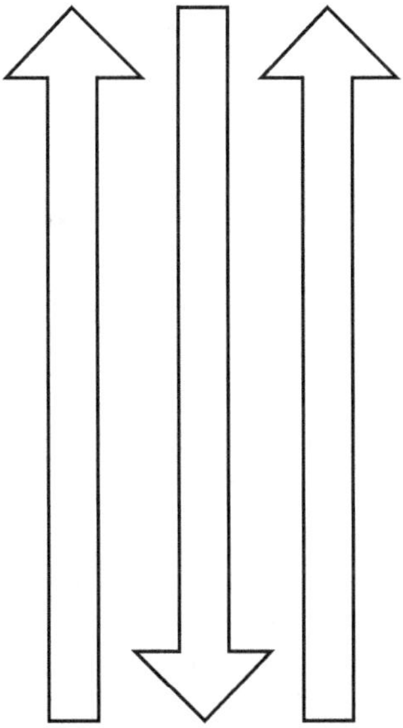

Fig. 3.9 A schematic representation of the beta sheet.

chains are associated. This is the quaternary structure. The reader may sense the complexity of this process by observing models of proteins. The reason for this complexity is that the protein folds and generates a three-dimensional structure from a linear sequence of amino acids. So far, no one has been able to completely determine the three-dimensional structure of the protein and hence its function by using the basic linear sequence. Why is it so difficult to conduct this reductionist move? After all, the protein is doing it all the time!

The existence of regularity in the protein's folding has led to what is known as "Levinthal paradox" although it is actually a problem rather than a logical paradox. The paradox is simple to explain. On the one hand it is assumed that the native state of the protein is thermodynamically the most stable state under biological conditions. In other words it is the state in which free energy is minimal. One may hypothesize that by using an *algorithm* the protein searches through a given "problem space" in order to find its energetically favored state. However this is impossible. The primary structure has zillions of possible conformations due to the enormous number of possible non-covalent bonds between the components of the polypeptides. For example, let as assume that each peptide group has only three possible conformations. A polypeptide chain of 150 residues would have $3^{150} = 10^{68}$ possible conformations! Even if we assume that the search time is minimal then it would take 10^{56} seconds to search all these conformations while it is known that the folding time *in vivo* and *in vitro* ranges from 0.1 to 1000 seconds (Branden and Tooze, 1999). That is, on the one hand, there is an impressive regularity in proteins' folding but, on the other hand, it is impossible to assume that the protein searches the whole range of possible conformations in order to achieve the most stable structure. Things are actually much more complicated. The change in the conformation of the protein is not only the result of internal interactions between the atoms but also a result of interaction with other molecules. There is no algorithm that leads us from sequence to structure. From a computational perspective this brute search is impossible (Finkelstein and Galzitskaya, 2004) yet computer scientists all over the world are competing in developing heuristics for predicting as much as possible the folding of proteins.

We may conclude by saying that the structure of the protein is related to its sequence. However, as in natural language what determines the meaning of an utterance is not the grammar *per se* but the actual, dynamic, and complex linguistic event that takes place in *context* and in *interaction* with another agent. Something similar is evident in the realm of living systems. Structure cannot be simply reduced to sequence; interaction in context is indispensable for understanding the behavior of living systems.

5. The Genetic System: Simple Mapping or a Complex Beehive?

During the division of the cell the two strands of DNA divorce from each other. Each strand is complementary to the other strand and therefore serves as a template for the synthesis of a complementary strand in the daughter cell. Therefore the *replication* process involves the production of two strands from one, a process that deals with huge numbers of replicated units with an astounding accuracy.

Let us turn back to the divorce of the two strands. As we know, in any process of divorce there is a third party, usually a smooth lawyer who makes his living from the separation process. The third party in the separation of the strands is a protein that breaks the hydrogen bonds that hold the strands. The places were this process of separation begins are called the *replication origins* and the human genome has approximately 10,000 origins. When those separating proteins start to work we can observe an unzipping of the two strands.

As in all stories we have a hero or more accurately heroes. Who are the heroes of the replication process? The heroes are a bunch of *enzymes* that function in a highly complex metabolic network for which DNA is only *raw data*. Understanding this complex metabolic network makes our simplified image of genetic activity, as a simple mapping from DNA to proteins, irrelevant and primitive. Let us explain the role of the enzymes in order to sense the complexity of this process.

The first hero is an enzyme named *DNA polymerase*. The name of an enzyme typically ends with *-ase*. Polymerase simply means that our enzyme is involved in catalyzing polymerization reactions in the synthesis of the DNA. It has the important role of synthesizing the new strand from the template by adding a new nucleotide to the growing chain.

Another enzyme—RNA polymerase—plays an important part in the process of transcription, the first step in switching on a gene, when the information carried in the DNA is copied onto a molecule called RNA. To start this process, the polymerase must first find a binding site known as a *promoter*, a fragment of DNA in front of the gene.

To see how the polymerase seeks out the promoter, researchers played a dirty trick (*The Economist*, 1997). The sequence of DNA they selected for the polymerase did not actually contain a promoter. When they came to watch the film produced by their special microscope, they saw the polymerase land on the DNA, and then slide up and down along it, jostled randomly in either direction by the thermal energy of the solution. From time to time, it would detach itself, and then settle somewhere else and start hunting again, alas in vain (*The Economist*, 1997). As we can see, the polymerase surfaces on the DNA by using thermal energy, we should also see that this enzyme is helpless without a contextual cue, a binding, which directs it where to start the process. Without contextual cues meaning can be generated neither in language nor in biology.

Not only is our enzyme involved in a construction process, it is also responsible for an error correcting activity—*proofreading*—that aims to control and remove mispaired nucleotides. Proofreading is a meta-cognitive or meta-biological activity. *Meta-* abilities urge us to examine both biological and cognitive systems as hierarchical systems that carefully regulate and control their own behavior. This observation has important implications for understanding living systems. These implications will be discussed in the following chapters.

Another enzyme—the *primase*—synthesizes short RNA primers during DNA replication. By using DNA as a template this enzyme creates a short length substitute, a closely related type of nucleic acid. This substitute—RNA (ribonucleic acid)—is similar to the DNA strand but contains the base U (uracil) instead of T, and ribonucleotide subunits in which the sugar is ribose. We need the RNA to initiate the replication process at the replication origin and to synthesize the RNA primer that initiates a new DNA strand.

In psychoanalytic terms we may describe RNA as a transitional object between the template and the new strand. Donald Winnicott coined the term *transitional object.* He argued that in the developmental phase in which the child separates himself from his mother (me from not-me) he uses certain objects, like his Teddy Bear, which support the separation process by providing a substitute for the not-me. As you can see a transitional object is not only important in psychology but in biology too and the reason is interesting. Development both in psychology and biology is a *mediated activity*. A direct interaction is evident only among particles of matter. Across scales of analysis, the realm of living systems is characterized by mediated activity. Later I will describe this form of mediation in semiotic terms as a sign-mediated activity (semiosis) but for now let us keep in mind that there is no *direct* transformation from DNA to protein. Mediation is a constituting principle of living systems. The major implication of this conclusion is clear: If biological activity is mediated then understanding the meaning and mechanisms of biological semiosis is a major challenge facing biological research. This challenge is far from being trivial since the meaning of sign-mediated activity is far from clear. In this book I propagate the thesis that biological systems are *sign-mediated* and delve into the meaning and implication of this thesis. Let us return to enzymes.

There are other enzymes that are involved in the process of transcription and translation. Enzymes that remove the RNA primer (nuclease), replace it with DNA (repair polymerase), and join the DNA pieces together (DNA ligase).

Let me add another level of complication to the book of life by describing other proteins involved in this process. *Helicase* is a protein that opens the double helix as it moves forward. A *single-strand binding protein* attaches to the strand released by the *helicase* and prevents it from re-forming base pairs. Another protein—*a sliding clamp*—attaches the DNA polymerase to the DNA template. And not a word has as yet been said about the translation process from the RNA to the protein. It should be noted however that each group of three consecutive nucleotides in the RNA comprises a *codon* that specifies one amino acid of the protein through the assistance of tRNA and aminoacyl (tRNA synthetases that assign the appropriate amino acid to its corresponding RNA molecule).

Our overall impression from the oversimplified description I have presented so far is that rather than a simple mapping from genes to proteins, the construction of the organism looks more like the activity of a busy beehive. This picture is quite different from the oversimplified image sometimes portrayed to high school students through the "central dogma" of genetics.

According to what is known as the central dogma of molecular biology genetic information flows from the DNA molecules to the RNA molecules (a process known as transcription) to the protein (a process known as translation). Presenting the flow from RNA to proteins as a process of translation has the benefit of simplifying the process and providing us with a nice metaphor. However, this metaphor is misleading in a certain sense. Is the flow from the RNA molecules to the proteins like a translation between two languages? Is translation a simple mapping between two alphabets? This suggestion does no justice to the complexity of genetics and natural language. Here is an autobiographical anecdote illustrating the fallacy of considering translation as a simple mapping at the tokens level.

As a child I decided to learn English by myself. I wrote the Hebrew alphabet in a list and asked my father to write the corresponding list of the English alphabet. I put the two lists near each other and did my best to find a one-to-one correspondence between the two alphabets. After a partial success in accomplishing my task I moved on to the lexical level. I took English words, replaced their tokens with their corresponding Hebrew letters and found, to my disappointment, that I produced gibberish. I failed at my mission but learned a lesson about translation. Translation does not exist at the token level. Translation always assumes context.

Another aspect of genetic system complexity concerns the active role of the genes. One may think of the genes as passive objects that are transcribed/translated from one form to another. Again, the metaphor is the one of transcribing a code from one language into a different language. This image is wrong. There is no clear one-to-one mapping from a gene to a phenotype and, it has been found, in some cases there is an interaction between the genes. This phenomenon is known as *Epistasis* and it describes a situation in which the differences in the phenotypic value of an allele (any one of a series of two or more different genes that occupy the same position, or locus, on a chromosome) at one locus are dependent on differences in specific alleles at one or more other loci (Wade *et al.*, 2001). Moreover, genes are not passive codes but some of them take an *active part* in mapping processes! There are regulatory DNA sequences that are needed in order to switch the genes off and on, and there are gene regulatory proteins that help them to express the genes. These regulatory genes are not passive letters in the book of life but active letters that are involved in reading the book of which they are a part. Therefore gene expression is not a simple process and

> Whether a gene is expressed or not depends on a variety of factors, including the type of the cell, its surroundings, its age, and extracellular signals. (Alberts *et al.*, 1998, p. 259)

This statement has radical meaning and one should notice that it appears in *Essential Cell Biology* and not in a post-modernist text. The meaning of this statement is that gene expression is a *context-dependent process*. Context is a key concept for understanding any meaning-making activity whether in human interaction or in the genetic system.

What have we learned from the discussion so far? A very simple lesson: It turns out that instead of a linear sequence of letters that are passively transcribed and translated by an external observer, instead of a linear sequence that determines in a simple causal manner the construction of the organism, we have a complex network of cooperating agents that are responsible for actively taking "raw genetic data" and turning it into a living organization of which they are a part. Things however are even more complex since this raw genetic data is more active than we originally believed. Rather than passive letters in a book the genes seem to be more like the active and opinionated Hebrew letters from which the world was created.

6. The Reader's Role

The realization that things are much more complex than we had assumed at the beginning should not surprise the intellectual. They are always more complex than we assumed at the beginning. One may find comfort in knowing that this is not a defect of genetic research but an awareness that characterizes the evolution of theories of interpretation in general.

The evolution of an interpretation process, whether of texts or of biological processes, is such that it proceeds from a simple and literal understanding of the text to the recognition of meaning as the reader's role, and from the reader's role to the Bakhtinian conception of the mutually constituting triad of author, reader, and hero (or text) immersed in an interconnected web of signs.[2] Let us explain this argument. Historically, *hermeneutics*—the study of interpretation—evolved from a literal understanding of the text as simply conveying meaning to an interpretive position toward the text. The Bible was considered to be a simple mirror of God's word and interpretation was the one and complete way to extract God's message from the holy text. Today, we call those folks who still hold this approach to interpretation as fundamentalists. However, our understanding of the interpretation process evolved in two important aspects.

First, we broadened our scope of what a text is. Text is not limited to the things written in the Holy Scriptures or even to things written, in general. Texts are a conglomerate of tokens that we interpret in a meaningful way, so cultural and biological texts are possible. Fridrich Nietzsche and Roland Barthes are just two of the intellectuals who enlarged our conception of what a text is.

---

[2] Evolution goes beyond this point to an even higher level of complexity that will be discussed in the concluding chapter.

Second, we shift our attention from the *text* and the *author* to the *reader*, from the meaning allegedly encapsulated in the text or from the author's intentions, to the process through which meaning is actively constructed by the reader from the raw data of the text (Barthes, 1977). In biological systems this interpretation is far more complex than we naively imagine because the text is not just raw data, but has its own active role in the reading process. Like Bakhtin's heroes, genes have something to say.

The reader may correctly identify the names of those who are associated with this idea, like Roland Barthes. However, it is more interesting to consider how this approach or *Zeitgeist* percolated into our understanding of biological processes. For example, years ago it was found that many of the genes in higher-order organisms are fragmented and composed of expressed segments of the DNA (*i.e.* exons) and intermingled with *junk DNA* known as *introns*. The finding of junk DNA created an enormous headache for those who hold the idea that there is a relatively simple transcript from DNA sequence to proteins. To make matters worse (or better) it was found that exons can be spliced in more than one way and that different mRNA transcripts (interpretations?) can be extracted from a single primary transcript (text?). This process, again, is mediated by enzymes (snRNPs) known as *snurps* that are complexes of proteins and RNA. Keller (2000) who discusses the meaning of these findings writes:

> The bottom line is that, depending on the context and stage of development of the organism in which a primary transcript find itself, different pieces of the transcript may be cut and pasted together to form a variety of new templates for the construction of a corresponding variety of proteins. (pp. 60–61)

Is this not a clear-cut case of a post-modernist approach to biological text? Does the fact that a single gene encodes not a single protein but many other proteins not correspond to the polysemy of signs in natural language, to the fact that a single word may have different meanings in different contexts? To a certain extent the answer to these questions is positive. However, any biological system presents regularity, so the post-modernist relativist slogan of "anything goes" cannot be applied to the biological realm. After all, human beings bring into the world other human beings, not dinosaurs.

7. Summary and Conclusions

In this part of the book, I tried to give the reader a sense of the complexity of the genetic system and the limits of classical reductionism in helping us to understand this complex system. We learned that a reductionist move is inevitable on the one hand but extremely limited on the other. After breaking the system into its components we must understand the way interactions in context result in the whole functioning we would like to understand. This naturally leads us to a semiotic perspective on living systems. Interactions in living systems are not direct mechanical encounters but events mediated by signs. This important idea establishes new grounds for studying living systems. If living systems are constituted through sign-mediated activity, then non-reductionist biologists should adopt a *bio-semiotic* perspective. Several scholars have proposed this idea, and I will use it as a general theoretical framework for my analysis.

The next two chapters aim to illustrate the benefits of adopting this framework by considering concrete issues in biology. Chapter 4, "Why are Organisms Irreducible", explains the need for semiotic mediation in biological systems and presents a novel explanation for the irreducibility of biological systems. The following Chapter 5, "Does the Genetic System Include a Meta-Language?" illustrates a semiotic perspective on genetics by explaining the function of non-codable RNA, or what has been known as junk DNA, in terms of meta-language.

Chapter 4

A Point for Thought: Why are Organisms Irreducible?

Summary

In the previous chapters I discussed the limits of reductionism and illustrated the limits of genetic reductionism. In this chapter, I present a novel argument for why organisms are irreducible. To present this argument, I begin by addressing a fundamental question: Why are there sign-mediated interactions in biology? According to Polanyi, biological hierarchies are constituted through boundary conditions. I argue that signs, or more accurately the processes of signification, function as these boundary conditions. Moreover, based on general insights from the physics of computation, I argue that the organism cannot be computed directly from DNA without the loss of critical information. In this context, signs as boundary conditions mediate biological construction in a way that prevents the loss of information and the destabilization of DNA.

1. Introduction

In the minds of scholars and laypersons, the concept of sign is usually associated with the linguistic realm in which signs are used as a vehicle of communication between human agents. However, in the most general sense, a sign can be considered to be a "carrier of meaning" and as such it exists in biology. The reason I put the expression "carrier of meaning" in quotation marks is that meaning cannot be carried as if it were an object. This is a misleading metaphor. Meaning always involves a response and is thus an activity rather than an object, an invitation for a dialogue rather the transformation of information. This idea is clearly presented by Holquist (1990a):

> Lack of water means nothing without the response of thirst ... It is still the case that nothing means anything until it achieves a response. (p. 48)

For example, the meaning of a monstrous face staring at us from the dark is the response of fear and/or flight. No meaning is encapsulated in the face. The meaning of a molecule classified as an antigen is comprehended only through the immune response (Cohen, 2000b). No meaning is encapsulated in the molecule. Why do I emphasize the idea of meaning as a response? For two reasons: first, to dismiss the misconception that meaning is encapsulated in the message and can be reduced to the message; and second, to introduce the idea that the sign is not a literal carrier of meaning but a trigger (or a cue) for meaning-making.

A variety of *triggers of meaning*, such as mRNA, cytokines, and hormones, are evident in organisms, specifically those that are considered to be higher-order in terms of various complexity measures. Those triggers of meaning are studied through biological, physical, or chemical lenses. However, as signs they may also be approached from a more general *biosemiotics* perspective that deals with issues of signs and signification in the biological realm. In this context, a semiotic analysis may at least provide theoretical biology with some interesting suggestions regarding signs and signification in living systems and with a novel argument for why organisms are irreducible.

The relevance of a semiotic analysis to biology was originally developed in the first half of the 20th century by Jacob von Uexküll (1982). His work made it clear not only that the concepts of sign and signification are relevant to theoretical biology but that they are indispensable for understanding the unique nature of living systems as opposed to matter. The aim of this chapter is not to propagate or to present this theoretical perspective, a task that has been done by others (*e.g.* Hoffmeyer, 1996; Sebeok, 2001), but first to address a fundamental question in theoretical biology from a biosemiotic perspective. The question is:

> Why are sign-mediated activities evident in organisms?

This question will be addressed specifically with regard to the genetic realm of higher-order organisms. The answer will hopefully provide us with an explanation for why organisms so fiercely resist a reductionist explanation. This answer addresses the difficulties mentioned in previous chapters and presents a prospective vision for the chapters to come.

2. Why Do We Need Signs?

Any inquiry into biology from a semiotic perspective should address the question of when and why a direct encounter between biological components/systems is possible and when and why sign-mediated interaction is a must. This question is far from mere philosophical casuistry; its implications are evident in biology. Certain interactions, specifically at the atomic and molecular level of analysis, do not seem to be sign-mediated. For example, the interaction of enzyme and substrate may be described according to the lock-and-key metaphor (Clardy, 1999) as a direct structural encounter between two entities through non-covalent forces. No signs are evident in this encounter. In other cases, such as the transformation from DNA to proteins, mediation through mRNA is clearly evident.

One may consider the question of why DNA cannot be used directly to synthesize proteins without the mediation of RNA. Answers to this question—or, more accurately, scholarly speculations—can be provided indirectly from functional or evolutionary perspectives. However, these answers do not directly address the question and the reason for the existence of semiotic mediation in the realm of living organisms. For example, Francis Crick argued that life (or at least genetic replication) started with RNA. DNA, which is a more stable molecule and is better for long-term storage of genetic information, came later. This argument concerns the evolutionary primacy of RNA but does not *explain* why RNA is a necessary mediator for protein synthesis.

Another explanation is that because there is usually only one copy of any particular gene in the cell, the movement from DNA to protein is much more rapidly mediated through RNA (Alberts *et al.*, 1998). In other words, RNA allows

> synthesizing the required amount of protein much more rapidly than if the DNA itself were acting as a direct template for protein synthesis. (p. 212)

Another possible explanation for the existence of RNA is a functional one. DNA cannot leave the nucleus. However, "information" must be carried out from the nucleus to the ribosome. In this sense mRNA clearly functions as a sign since it functions as a "carrier of information" from one system to the other. These answers are either evolutionary or functionally oriented. However, there are other perspectives and in the following sections I would like to provide my explanation from a semiotic perspective. To address this question from a semiotic perspective, I would like to discuss the term *transmutation*.

3. Life's Irreducible Structure

In his seminal paper on translation, the linguist Roman Jakobson (1971) made an important distinction between two concepts: *translation* and *transmutation*. Translation concerns a transformation from one semiotic system (*i.e.* a system of signs) to another semiotic system, and transmutation involves transformation from a verbal to a non-verbal semiotic system. The term *transmutation* may be enlarged to include transformation between non-semiotic systems through semiotic mediation, and here I will use it in this specific sense.

Why is the concept of transmutation relevant to our inquiry? To explain its relevance, we should move on to Michael Polanyi's (1968) classic paper "Life's Irreducible Structure".

As a scientist, Michael Polanyi did not doubt that organisms are composed of matter that may be described in physical terms. However, the main point of his paper is that what is important for our understanding of living systems is not matter as such but the structure of boundary conditions or the restrictions that constitute the biological hierarchies of which organisms are composed. To use an analogy:

> When a sculptor shapes a stone and a painter composes a painting, our interest lies in the *boundaries* imposed on a material, and not in the material itself. (Polanyi, 1968, p. 1308; emphasis mine)

Please keep the sculpture metaphor in your mind. As suggested by Michelangelo, constructing a sculpture (or an organism) involves throwing things away, not just positive addition. Loss, oblivion, and irreversibility as major themes in the organism's construction and maintenance will appear again and again in this book.

The most important implication of Polanyi's argument for the current chapter is as follows: The biological hierarchy is composed of various levels that can be described in physical terms. If these levels interact through boundary conditions to constitute the living organism, then the interface between one level and another within the biological hierarchy is far from trivial. This interface cannot be simply reduced to the laws governing the levels themselves. It is a transformation from one form of matter to another form of matter. This transformation must be mediated according to rules that cannot be reduced to the laws of physics.

> *My first suggestion is that semiosis is the activity that enables the shift from one form of organized matter into another form of organized matter.*

In other words, in biology, *semiosis*—sign-mediated interaction—is a "vehicle" for moving between subsystems and levels and it serves as a boundary condition (Polanyi, 1968) for the transformation between different layers of matter (*i.e.* non-semiotic systems). Living matter is mediated and therefore constituted by signs or more accurately the dynamic work of signs (semiosis).

Moreover, the biological hierarchies discussed by Polanyi may be considered the "output" of a computation process performed on the "input", the genome. For instructional reasons and for the current phase of our inquiry I present organisms as being computed from the genome. This presentation is oversimplified since organisms are not like the computers we know and more will be said about that throughout the book. Later I will use computation in a different, wider sense and explain that organisms are *recursive-hierarchical* "machines" that compute themselves. In the meantime let us stick to the popular notion of the organism as computed from the genome.

According to this idea, which is the bread and butter of modern biology, a computational process should be understood in its most general sense as a procedure that produces an output from a given input. However, according to insights gained in the physics of computation (Landauer and Bennett, 1985), any process of computation involves the loss of information. This idea will be presented and elaborated later, but for the time being let us adopt it as is. If we accept the idea that computation involves the loss of information then what happens to an organism that computes itself from the genome? *The answer is if higher levels of the organism's hierarchy are the computational output of lower-level DNA, then the result is inevitably a loss of information in the transformation between the levels.*

> *Semiosis functions not only as the boundary condition but as unique interface that mediates the computation of the genomic input without causing a loss of information.*

To sum up, my main argument is that a major function of semiosis, at least in higher-order organisms, is to mediate between different layers/systems of biological hierarchies and to compensate for the irreversible process of computation evident in the construction and maintenance of the organism. This idea is elaborated upon in the next sections.

4. Transcending the View "From Within"

Let me begin my discussion by asking a very general question: What is the minimum condition for semiosis, that is, for the use of signs in a system? The answer is clear: *a difference*. At the heart of any semiotic activity, we must assume the existence of differentiated states (*e.g.* genes, letters). In a world of unity no signs are evident and no semiosis takes place. In the Book of

Genesis, for instance, differentiated states (fowls, beasts, etc.) were created first, and only afterwards did Adam name them by using signs.

If we accept the idea that differences are the most primitive ontological units, then those differentiated states may be mistakenly considered as physical distinctions, which are the minimum units of a semiotic analysis. This idea represents the stance known as naïve *realism* that assumes our mind is preceded by a reality expressed in physical terms. A semiotician cannot be a naïve realist. For the semiotician life is always mediated and what Kant described as the *Ding-an-Sich*—the thing in itself—is no more than the projection of our naïve fantasies on the world of signification.

An immediate response to the mediated conception of the world is that the existence of differentiated physical states at the base of our ontological hierarchy does not necessarily imply that there is a contemplating mind interpreting these differentiated states as signs. After all, differentiated physical states existed long before organisms started populating the earth. Well, this is of course not a new argument and the antagonist to my argument may pull his second gun by ironically asking me whether Newton's laws did not exist before Newton formulated them.

The fight between naïve realists and constructivists has a long and a rather boring history. I have no intention to discuss it here. My only argument is that a physical distinction, whatever it is, cannot be truly used as a unit of signification. The fact that physical distinctions are distinctions as long as they are being identified by a certain mind is the theoretical position I adopt. Both the naïve realist and the "naïve semiotician" are right. Any physical difference/distinction is a singular event without the existence of a contemplating device that may convert it into a more general instance of a class. Deleuze (1990) even coined the term *repetition* to describe this ontological category of a "difference without a concept", or, to borrow from phenomenological jargon, a difference "in and for itself". Deleuze's highly abstract idea of difference and repetition as constituting reality will be discussed in the concluding chapter. Meanwhile, let us adopt the idea of a difference without a concept as constituting the basic level of our ontology. Let me suggest that the realm of the living is layered in three levels:

Level 1: A Repetition. A difference without a concept.
Level 2: A Concept. A difference that makes a difference/functional generality.
Level 3: A Sign. Third-order differentiation. Communicated functional generality.

Let me explain this hierarchy. Physical states are pure singularities, a property derived from the fact that they occupy different positions in space-time. For example, as an organism grounded in a physical reality, each cat is a unique creature with its particular position in space and time. It is what Peirce described as the *dynamical object*. When we experience the world, on the most basic level we encounter these singularities (*i.e.* repetitions) that result from a basic encounter between the mind and the world. At the next level concepts emerge. Only the ability of a contemplating mind to *group* the various instances of the cat (*i.e.* the repetitions) under the concept *Cat* makes it possible to approach the singular cat from a general perspective and to communicate this general perspective with other human beings. In other words, instead of accepting the dichotomy between mind and nature as a starting point, I adopt the interactionist perspective as presented in my previous book (Neuman, 2003a). In this context, I would like to define a sign as follows:

A sign is a functional generality that is communicated across realms.

Let me illustrate this idea. Each of the four DNA nucleotides is a singularity (level 1), since it occupies a unique position in the linear sequence of DNA. However, when transcribed into mRNA these nucleotides lose their singularity to become a part of the DNA triplets known as codons, each signifying/specifying the synthesis of a specific type of an amino acid. In itself a codon is a concept: a second-order difference, a "difference that makes a difference", a functional generality. When communicated to determine a protein it turns into a sign. A codon is a sign—a communicated functional generality. For example, in the standard code, the codon CUC specifies leucine, the codon CGU specifies arginine, and the codon CAU specifies histidine.

The codons are signs in another sense as well: in the sense that they may signify different things to different observers. Although in the standard code CUU specifies leucine, for yeast it specifies threonine! In other words, there is no one-to-one correspondence between the sign and the response it invites (*i.e.* its meaning). This is a general characteristic of signs, and biological signs are no exception.

Codons are also signs in another important sense. A sign may turn out to be a signified realm (*i.e.* a source of signification) itself. Codons may also turn out to be signified realms themselves. To reiterate, codons are used for the synthesis of amino acids through the mediation of transfer RNA (tRNA). Through the anticodon region, tRNA adheres to the codon, and through a short single-strand region at the 3′ end of the molecule, the amino acid that matches the codon is attached to the tRNA. In this case, it is the tRNA that turns out to be a signifying process.

Now, we start to understand why signs as communicated functional generalities are necessary in biological systems. As argued by Barbieri (2005):

> All inorganic molecules are made by self assembly and their structure is determined *from within, i.e.* by internal factors. (p. 115)

In contrast, genes and proteins are produced by molecular machines,

> which physically stick their subunits together in an order provided *from without*, by *external templates*. (Barbieri, 2005, p. 115)

Following Barbieri, we understand that physical systems, as conglomerates of singularities, are embodied in local interactions from *within* the system, while biological systems are capable of synthesizing proteins because they can transcend the "view from within" through the generality and the communicability of signs/codes. Indeed, transcending the view from within is a constitutive dynamic of living systems that is possible only through signification. Let us move on to our next station as we continue our journey.

5. Information as a "Difference That Makes a Difference"

Gregory Bateson was the son of the distinguished geneticist William Bateson who named his son Gregory after Gregor Mandel. Bateson was a polymath whose work produced insights into a variety of domains, including family therapy and theoretical biology. The work of Bateson contains a wealth of significant ideas, relevant for understanding the biological realm. One of Bateson's significant ideas concerns *difference* as the organism's basic unit of analysis.

As Bateson (2000) argued in his seminal essay "Form, Substance and Difference" (originally published in 1970), the realm of the living is a realm in which effects (*i.e.* responses) are brought about by information, which he uniquely defines as a "difference that makes a difference". That is, in the realm of the living, pure differences are not enough. Only differentiated states (*i.e.* differences) that are actively differentiated on a higher level of analysis (*i.e.* a difference that makes a difference) can constitute the realm of the living. Bateson's idea of a difference that makes a difference will be repeated in this book again and again and again and ... again. It is a constituting idea of this book and as such deserves the appropriate redundancy.

One important implication of Bateson's idea concerns the notion of the *observer*. For instance, one should notice that the existence of an observer is built into Bateson's definition of information. A difference may make a difference only to something or someone (*e.g.* molecular machinery). Remember the genetic book of life and the question, "who is reading the book of life?" Bateson's idea has the potential for addressing this question by suggesting that organisms are unique books that read themselves.

Why do we need Bateson and Polanyi's ideas to understand sign-mediated activity? One answer I would like to give is somehow surprising, although it is rooted in an idea that appears in a slightly different form in Polanyi's paper. Based on Barbieri and Bateson, we may suggest that the realm of the living is not simply a realm of complex combinations of particles of matter. It is a realm in which matter is *actively* transformed by molecular devices into informational content (*i.e.* a boundary condition or a difference that makes a difference), which is recursively used for the construction and constitution of the organism. In other words:

> *An organism is constituted as a recursive hierarchy that through semiosis turns physical singularities into informational content for the active production of other physical singularities (i.e. its self-creation, or autopoiesis).*

That is, novelty of biological construction through informational content is the living system's *sine qua non.* However, the physics of computation teaches us a lesson about the price of this process, constraints that should be taken into consideration through sign-mediated activity. This lesson will add another layer to our understanding of semiosis in living systems.

6. Machines of Oblivion

Novelty, as is evident in the realm of the living, results not only from turning physical differences into informational content, but also from *erasing information* in order to create qualitatively new structures. That is, at the heart of emerging biological structures is the idea that something is necessarily gained and something is necessarily lost when we shift between levels of the biological hierarchy. This state of affairs is crystal clear for the embryologist who observes the apoptosis of cells as a natural process of embryonic development.

Jorge Luis Borges (2000a, p. 183) beautifully epitomizes the close relation between novelty and oblivion in one of his stories:

> Solomon saith: *There is no new thing upon the earth.*

> So that as Plato had an imagination, *that all knowledge was but remembrance*; so Solomon giveth his sentence, that all novelty is but oblivion.

As Borges, who attributes the above piece to Francis Bacon, suggests, "all novelty is but oblivion". This idea might seem too poetic for biologists but it may become comprehensible when examined in terms of the physics of computation. Indeed, a recent paper on the physics of computation was entitled "The Physics of Forgetting" (Plenio and Vitelli, 2001).

Let us return to the argument I previously mentioned regarding computation and the loss of information. As argued by Landauer and Bennett (1985), a process of computation involves the loss of information. They define computation as a process in which an *output* is produced from an *input*, and *information* is considered in the most general sense of differentiated states.

Let me explain the loss of information with a simple example: The arithmetic expression $1+1=2$ involves a process of computation in which the binary operation of adding the inputs 1 and 1 produces the output 2. Why and in what sense does this process involve the loss of information? There are physical and computational ways of describing and explaining this loss.

One way to think of information erasure is in terms of computational irreversibility. A logical gate such as OR *is irreversible if, given the output of the gate, the input is not uniquely determined.* For example, the logical gate NAND (not and) is intrinsically irreversible. If the output of the gate is 1 then the input could have been 00, 01, or 10 (Nielsen and Chuang, 2000). The same process is evident with regard to the simple arithmetic expression previously presented. Without getting into the particularities, qualifications, and difficulties of this argument, we should acknowledge the common-sensical idea that computation involves the loss of information in the most general sense of *differentiated states* (Landauer and Bennett, 1985). In other words, when a difference is turned into a difference that makes a difference, some information/differentiation that exists at a lower level of analysis is lost.

The same idea holds in biology. Each cell in our body contains the same DNA. However, the computation of the biological hierarchy from this information source clearly involves a loss of differentiation at the different stationary states of the construction process, as is evident in the transformation from DNA to RNA. Here, we get into the issue of reductionism and why it is limited as a scientific explanation.

7. Reductionism is Impossible Because We are Irreversible

The idea presented above has crucial implications for understanding reductionism in biology from a novel perspective. Let me explain this argument: At this point in our discussion it is clear that the fact that higher-order organisms do not easily lend themselves to reducible analysis is not due to the failure of brilliant minds to accomplish the task. It is an intrinsic property of multi-level biological systems that are clearly irreducible due to the irreversibility of their construction process. In other words, *the construction of the organism is an irreversible process of computation* and therefore simply tracing the biological output back to its constitutive elements is impossible. Information is necessarily lost in the process of constructing the organism and this loss is the *sine qua non* of the construction process.

The question is: So what? Even if the construction of the organism involves a certain loss of information, as implied by the physics of computation, what has semiosis got to do with it?

8. The Digital and the Analogical

In my presentation above, *I was deliberately misleading*. Living systems are not typical computational machines. They are irreversible in the sense that genomic information is necessarily lost in the computation of the biological hierarchy. However—and this is the important point—they are computational machines that jealously preserve their input (DNA) in every cell of their bodies.

The most important implication of this observation is that a direct computation from DNA would have resulted in a loss of information and the diminution of the organism's most important "text" of stability. In other words, a direct, unmediated computation from the DNA would have been incompatible with the essence of the DNA as a source of informational stability. The solution is *sign-mediated computation*. DNA, a relatively stable source of information, remains intact as an input; it is only copied according to a one-to-one correspondence that preserves its identity and its informational content. On the other hand, the genome is continuously interpreted through sign-mediated activity that leads to the constitution of the organism.

Readers familiar with biosemiotics literature may immediately associate this thesis with the *code duality* of Hoffmeyer and Emmeche (1991). However, long before code duality was introduced, Gregory Bateson made the clear-cut distinction between *digital* and *analogue* modes of communication. This idea will also be discussed again and again through the book and I hope that the reader will not be bored by this intellectual rumination. These two modes of communication are critical for understanding the realm of the living.

A digital code involves a string of discrete tokens. It is a Turing machine style tape. After carefully reading Bateson, I believe that the most important feature of the digital code is that it operates *on the same level of logical analysis* (Bateson, 2000, p. 291). Let me explain.

In Bateson's terms (2000, pp. 140, 291), DNA is a digital code of communication. This is why the only direct operation in which DNA is involved is *copying*, an operation that takes place as a one-to-one mapping on the same logical level of analysis. Copying preserves the DNA's digital code: guanine is guanine is guanine! Copying, however, can create nothing new. To construct an organism we must transcend the view from within, we must transcend the digital code, and therefore the analogue code that concerns magnitude, quantity, and similarity must be used. However, any form of coding directly from the DNA would have resulted in the erasure of the informational input. Remember the lesson we learned from the physics of computation? You cannot beat City Hall! Nature's solution is the sign. The sign is a functional generality that does not threaten the stability of the DNA. The information represented by the DNA is not erased. It is *interpreted* in the most basic sense of the term. The use of the term *interpretation* in the biological sense is not an intellectual whim. Biological interpretation, like linguistic interpretation, is a sign-mediated process. I use the term *interpretation* to denote

> *the generation of different macro structures from the same set of distinctive tokens.*

RNA codons are the result of an interpretation process. RNA uses the text as a point of reference (*i.e.* as an input) but does not dismiss it. It is a unique form of computation in which the stability of the DNA is assured and at the same time transcendent in a way that allows for the construction of novel biological forms.

9. Conclusions

Organisms are machines of novelty and oblivion. Through boundary conditions they transcend the view from within and constitute the living organism as sign-mediated matter. However, living systems have to strike a delicate balance between novelty and oblivion. Remember the dedication at the beginning of the book? Remember the logic of *in between*? This is a specific instance of this logic and many other instances will be presented in this book.

The ideal of throwing out the past in favor of new foundations is epitomized in the old Communist hymn "The Internationale". In the past, Billions of people around the world sang, "The earth shall rise on new foundations". Organisms do not accept this old Communist ideal; they retain their genes—their old "foundations"—as much as possible. So strong is this tendency to preserve the holy scriptures of the genome that a whole theory, that of the *selfish gene*, was constructed around this observation (Dawkins, 1976). However, organisms do not accept the fundamentalist idea of rejecting the present in favor of ancient holy texts either. Organisms are not the slaves of their genes just as they are not their masters. Like Talmudic sages, organisms interpret the ancient text of the genes for their autopoiesis and survival in the present. They are hermeneutic machines (Markos, 2002) that materialize the logic of in between.

This process is far from being simple and simplistic models of textual interpretation would not capture its real spirit. Reading the distant and ancient text of the genes creates a problem for the organism, which is solved by *dual coding* (Bateson, 2000). Surprisingly as it may sound, this problem is encountered by philologists, too.

The Spanish philologist José Ortega y Gasset (1959, quoted in Becker, 2000, p. 371) began one of his seminars by discussing the difficulty of reading. "To read a distant text", he wrote, "distant in space, time, or conceptual world—is a utopian task". This task is no different from the task facing an organism that is reading the distant text of its genes. The task, adds Ortega y Gasset, is one whose

> initial intention cannot be fulfilled in the development of its activity and which has to be satisfied with approximations essentially contradictory to the purpose which had started it. (Gasset, 1959, quoted in Becker, 2000, p. 371)

Commenting on this statement, the linguistic anthropologist Anton Becker says:

> In that sense the activity of language is in many particular ways utopian: One can never convey what one wants to convey. (Becker, 2000, p. 298)

As Ortega y Gasset puts it, it is *deficient* in the sense that it says less than it wishes to say, and it is *exuberant* in the sense that "it says more than it plans" (Becker, 2000, p. 298). This utopian characteristic of language is a source of flexibility that results from signs that are simultaneously deficient and exuberant. A sign always says less than it plans in the sense that as a functional generality it may serve different functions in different contexts. It is exuberant and always says something it did not plan in the sense of a generality that transcends the level of logical analysis from which it emerges. Codons are simple instances of these properties, but this chapter does not exhaust the list of sign activities in biological systems or the complexity, depth, and mystery of sign-mediated activities. The next chapter aims to move along the same line and to add another layer of complexity to our understanding of the semiosis that constitutes living forms. If you thought that a meaning-making perspective can be exhausted by a simplistic textual metaphor it is best to think again. More thought challenging ideas are waiting just around the corner.

Chapter 5

A Point for Thought: Does the Genetic System Include a Meta-Language?

Summary

In this chapter I aim to add another layer of complexity to our semiotic understanding of the genetic system and the poverty of reductionism. The issue I have chosen is the one of *junk DNA*. Non-codable DNA sequences were described as non-functional junk DNA. However, more and more evidence is being gathered about the different functions fulfilled by ncRNAs. In this chapter, I wish to consider *ncRNAs as a part of a Meta-language*. More specifically, I argue that every language or more generally, every system of signification must have a complementary meta-language (or a meta-system) for its functioning. In this context, the genetic realm is not an exception and the genetic "language" must be accompanied by a meta-language, which is (partially) materialized by the ncRNAs.

1. Introduction

In the introductory chapters, I presented the oversimplistic and mechanistic dogma of genetics and tried to point to its shortcomings. The central dogma of genetics propagated through the mediation of the linguistic metaphor (Alberts *et al.*, 1998). In this context, the transformation from the DNA to the RNA has been described as a process of transcription and the transformation from the RNA to the proteins has been described as a process of translation.

Metaphors are indispensable in the realm of science the same as they are in any expression of thinking (Lakoff and Johnson, 1999). However, the role of metaphors in science is restricted and should not be confused with the role of a scientific model. As Tauber (1996) suggests:

> Theory must grope for its footing in common experience and language. By its very nature the metaphor *evokes and suggests* but cannot precisely detail the phenomena of concern. (p. 18; emphasis mine)

Having the role of evoking thought, metaphors are powerful and indispensable tools that may open new horizons for research. As argued by Efroni and Cohen (2003) a good biological theory is one that serves the process of discovery and opens the way to "otherwise unthinkable research". I like this idea because it emphasizes the creative and open-ended nature of scientific inquiry. This idea also points to the importance of metaphors in scientific discovery. Good metaphors are sometimes our gate to unthinkable research.

Following Tauber, a metaphor has a significant role in research as evoking unthinkable research. However, if metaphors evoke, and open the way to unthinkable research they may also have the power to block unthinkable research or to distort our understanding of current findings! I believe that exactly this kind of distortion is evident when we use an oversimplified version of the linguistic metaphor in genetics and forget that language (or any process of semiosis) is always accompanied by a meta-language. Enlarging the linguistic metaphor to include meta-linguistic processes may help us to conceptualize and understand unresolved issues in genetic research such as the one of junk DNA.

To date, genetic research has not yet used all the possible meanings and nuances of the linguistic metaphor in order to explore the complexities of the biological realm. This state of affairs is unfortunate since the oversimplified use of the linguistic metaphor in genetics may hinder our understanding of genetic phenomena while, on the other hand, maintaining the central dogma that does not seem to represent the genetic processes in all of its complexity (Mattick, 2003). This argument does not aim to dismiss the importance of the linguistic metaphor in biology but to critically examine its use and to enlarge its scope for the working scientist. As will be later illustrated, the common use of metaphors in biology adoptes a misleading approach to the nature of a metaphor. In this chapter, I do not aim to dwell on this issue and my reference to the notion of metaphors in biology is rather general and commonsensical. However, by the end of the chapter the reader may find that the linguistic metaphor in genetics is not really a metaphor and that the processes I "metaphorically" describe as meta-linguistic is the way things actually work both in natural language and in the genetic system.

2. "Junk" DNA: Is It Really Junk?

There are five major types of DNA in the human genome (Wagner *et al.*, 1993):

1. Transcribed and translated;
2. Transcribed but not translated;

3. Not translated;
4. Not transcribed with a unique structure; and
5. Not transcribed with a repetitive structure.

Mice and human beings share approximately 99% of their protein coding genes (Mattick and Gagen, 2001). This allegedly minor difference is a continuous source of amusement for the general public. The similarity of human beings and mice does not dismiss the qualitative difference between human beings and mice. In higher-order organisms only a minority of the genetic transcripts code for genes. Therefore, the superiority of complex organisms is to be found elsewhere (Mattick, 2003) as will be described below.

The non-codable DNA sequences were described as junk DNA. What is the common explanation to the existence of this junk? A common answer is that those elements "do not have any function: They are simply useless, selfish DNA sequences that proliferate in our genome, making as many copies as possible" (Makalowski, 2003, p. 1246). This common explanation may be described as the "appendix explanation". Junk DNA is considered to be a non-functional and redundant remnant of our evolutionary heritage, the same as our appendix.

This "appendix explanation" appears in the authoritative *Essential cell biology* of Alberts and his colleagues (1998), a textbook that is one of the major sources for educating biologists. Alberts *et al.* (1998) elaborates on the "appendix explanation" by comparing complex organisms to bacteria and unicellular eukaryotes that do not have the same huge proportion of junk DNA. By comparing simple to complex organisms, it is argued by these authors that bacteria and simple unicellular eukaryotes are under strong selective pressure to divide at the maximum rate permitted by nutrients in the environment and thus to minimize the amount of superfluous DNA in their genome, as DNA replication is *costly in terms of energy* and material resources (Alberts *et al.*, 1998). In contrast to larger cells in multi-cellular organisms such considerations are less relevant and therefore there is no strong selective pressure to eliminate non-essential DNA sequences. This is the explanation for the huge proportion of junk DNA in the human genome.

It seems that this explanation which is "energy" laden can be questioned on a theoretical and empirical bases alike. Let me explain my objections by using an insight from a field seldom discussed by biologists: the physics of computation. The physics of computation was discussed in the previous chapter and its general insights will enrich the discussion through the entire book.

3. Is Keeping the Junk "Energetically Favorable" to Deletion?

As was argued by Landauer (1961) the elimination of information from a given system is an activity that consumes energy and dissipates heat into the environment:

> When an information is erased there is always an energy cost of kT ln 2 per classical bit to be paid … [and an] amount of heat equal to kT ln 2 is dumped in the environment at the end of the process. (Plenio and Vitelli, 2001, p. 27)

Considering biological systems in general computational terms, this argument should be taken into account whenever the issue of information deletion gets into our discussion.

Landauer's argument is thought provoking for two reasons. The first reason is that it associates the abstract mathematical term *information* with its commonsensical meaning of a differentiated realm and with the physical (and the bio-physical) realm. The second reason is that it associates the loss of information (in the general sense of differentiated states) with the release of heat to the environment.

The association between the dissipation of heat and the loss of information can be easily illustrated by using Landauer and Bennett's original example. Let us assume that we drop two identical elastic rubber balls from different heights: one meter and ten meters. The potential energy of the balls turns into the kinetic energy of movement. When the balls hit the ground they jump back and the height of their jump indicates the height from which they were dropped. As a note let me add that this example used by Landauer and Bennett associates information with measurement and observation, a statement which is not trivial from the perspective of Information Theory. As Bateson realized a long time ago, information and meaning cannot be dissociated from a contemplating mind whether the mind of a human being or the mind of the eco-system. The physics of computation implicitly accepts this opinion and the idea of meaning making as closely associated with measurement will be discussed later in the book. Back to our example: Whenever a ball hits the ground some amount of energy is being lost and we say that heat (*i.e.* energy in transfer) was released into the environment. After a while the two balls will rest peacefully on our playground indicating nothing about the height from which they were dropped. Heat was released and information was lost. In this sense, heat is not only the graveyard of energy that could have done some work (Hewitt, 1993) but the graveyard of information too.

In physics the efficiency of a system is defined in terms of the ratio between the energy invested and the work done. If all the energy was used to do the work then our system is perfect. However, if some energy was released into the environment then we are less than perfect. From the second law of thermodynamics we understand that no one is perfect. There is no system that can turn all its available energy into work. However, biological systems are highly efficient in their heat management when compared with man-made systems, such as the engines of our cars.

In this context, the elimination of biological information and the dissipation of heat into the environment is not a simple economical process of saving the energy of information copying as argued by Alberts *et al.* (1998). Genetic information does not simply fade away the way some aliens in science fiction movies do. The elimination of DNA sequences (*i.e.* biological information) is not a simple economic matter. There is logic behind those processes, a logic that is materialized through specific biological mechanisms that consume energy to do the elimination work. In this context the life and death "decision" about what kind of sequences should be removed as a result of evolutionary pressure may be just as energy consuming as the copy and storage of the "superfluous" genetic sequences. In fact, Landauer even argued that in contrast to the elimination of information *copying* classical information can be done reversibly, and (potentially) without wasting any energy!

I used basic ideas from the physics of computation to show that the energy-based explanation used by Alberts *et al.* is internally inconsistent. The attempt to explain the huge proportion of junk DNA in complex organisms by turning to energy calculations of evolutionary processes is internally inconsistent and scientifically shaky. So what is the explanation for the existence of junk DNA? First, let us realize as argued by Kidwell and Lisch (2001) that

> the selfish and junk DNA concepts have often been accepted blindly and rigidly to the exclusion of other host-elements relationships. (p. 1)

Selfish genes are an explanatory concept and there are other perspectives. In the following sections, I would like to explain the function of junk DNA by introducing recent research findings concerning non-codable RNAs and by introducing the idea of *ncRNAs as a part of a Meta-language.* More specifically, I argue that based on general semiotic principles every language or, more generally, every system of signs must have a complementary meta-language in order to function. In this context, the genetic realm is not an exception and genetic "language" must be accompanied by a meta-language, which is (partially) materialized by the ncRNAs. Therefore, my thesis shifts between abstract principles of semiotic systems and our knowledge and speculations with regard to ncRNAs.

4. ncRNAs as a Meta-Language

Mattick (2003) reviews the evidence that ncRNAs derived from introns of protein coding genes and the introns and the exons of non-protein-coding genes constitute the majority of the genomic programming in higher organisms. These RNAs are also described as functional RNA (fRNA) and includes different classes such as miRNA and snoRNA. These RNAs were considered of uncertain significance and have been studied only recently partly due to technical difficulties in studying these molecules and their function (Mattick, 2003).

Knowledge of ncRNAs has been limited to "biochemically abundant and anecdotal discoveries" (Eddy, 2001, p. 695). Nevertheless it was found that ncRNAs are involved in important biological processes and evidence in favor of this argument continues to accumulate. For example, it has been argued that ncRNAs may regulate protein synthesis by decelerating or accelerating mRNA degradation (Couzin, 2002; Voinnet, 2002). Another example of the ncRNA functions is splicing. It has been shown that introns are removed from the primary transcript of the RNA by enzymes that are composed of a complex protein and RNA. These splicing enzymes are called *Sunrps* (snRNPs). snRNPs are clearly involved in "meta-language" work. They are not the message itself but a tool for regulating the content of the message that is delivered through the mRNA. Silencing is another activity in which the ncRNAs are involved. Silencing is a classical example of the meta-linguistic nature of ncRNAs. Silencing is a meta-linguistic activity and the issue of silence (When? Why? Where?) is of great interest to linguists who are interested in the pragmatics of language (Jaworski, 1997). After all, it is common wisdom that "life and death are in the power of the tongue" and the realm of the genome should be no exception. Certain things should not be expressed or silenced either in human language or in genetic language. The issue of silence will occupy me throughout the book and the reader can expect to encounter it repeatedly.

It has been found that a variety of processes are affected by ncRNAs including transcription, gene silencing, replication, RNA processing, RNA modification, RNA stability, mRNA translation, protein stability, and protein translocation (Storz, 2002). Based on these findings, Mattick (2001) argued that:

> Phenotypic variation between both individuals and species may be based largely on differences in non-protein-coding sequences and

> be mainly a matter of variation in gene expression, *i.e.* due to the control architecture of the system ... (p. 986)

and that

> ncRNAs may constitute an endogenous control system that regulates the programmed pattern of gene expression during their development. (Mattick, 2003, p. 936)

In other words, sexologists are right when they argue that size is not as important as one would tend to believe! In our case it is not the size of the codable genome *per se* which is the "difference that makes a difference". What is important is what we do with it.

The importance of ncRNAs is further elaborated upon by Mattick through a general theoretical framework that challenges the simplicity and linearity of the central dogma. It is argued by Mattick that complex organisms require two levels of "programming". One level deals with the specification of the "functional components of the systems" mainly proteins, and the other level is responsible for the "orchestration of the expression and assembly of these components" (p. 930). Mattick argues further that the ncRNAs are involved in the second level of processing. They are involved in controlling and regulating genetic programming. In other words, junk DNA is not junk after all. A part of it (and this is my interpretation) is the basis for the *meta-language*, which is necessary and complementary to the language itself.

To explain this idea, I now turn to semiotics. This will allow me to explain the need for meta-language in any language, including the genetic one.

5. The Map and the Territory

To explain the unbreakable link between language and meta-language, I consider language in the most general sense and discuss the general relation between a sign and the signified. The relation between a sign and a signified is an intricate matter that can be approached from a more general perspective: the relation between a representation and the thing it represents. This delicate relation between the representation and the represented is insightfully illustrated in one of J. L. Borges stories: "On Exactitude in Science". As you can see I am a great fan of Borges and his stories are a continuous source of inspiration for my research and for illustrating my ideas.

In his story Borges describes an imaginary kingdom in which the art of cartography (the art of creating maps which is an art of signifying by itself) has reached a high degree of precision that allows the cartographers to create a map (*i.e.* a sign) that is the mirror image of reality (*i.e.* the signified). The end of this heroic venture is tragic:

> In time, those unconscionable maps no longer satisfied, and the cartographers' guild drew a map of the empire whose size was that of the empire, coinciding point for point with it. The following generations, who were not so fond of the study of cartography saw the vast map to be useless and permitted it to decay and fray under the sun and winters. In the deserts of the west, still today, there are tattered ruins of the map, inhabited by animals and beggars; and in all the land there is no other relic of the disciplines of geography. (Borges, 2000a, p. 325)

What is the lesson we can learn from this insightful story? The lesson is that by trying to create a representation of reality that turns out to be reality in itself the sign/map/code loses its unique power to signify. In this sense *any sign must maintain an unbridgeable gap between itself and the realm it signifies.* This important conclusion will be used in the final chapter of the book to explain signification as grounded in the dimensionality reduction that necessarily accompanies the representation of the world by the organism.

The relation between the sign and the signified was also discussed by Gregory Bateson in similar terms as the relation between a map and the territory it signifies. In one of his seminal papers "Form, Substance and Difference", Bateson points to the essential impossibility of knowing what the territory really is, as any understanding of it is based on some representation:

> We say the map is different from the territory. But what is the territory? Operationally, somebody went out with a retina or a measuring stick and made representations which were then put on paper. What is on the paper map is a representation of what was in the retinal representation of the man who made the map; and as you push the question back, what you find is an infinite regress, an infinite series of maps. The territory never gets in at all. ... Always, the process of representation will filter it out so that the mental world is only maps of maps, ad infinitum. (Bateson, 2000, p. 460)

Bateson's idea of the mental world as "maps of maps" should not be taken at face value as leading to infinite regression of maps. Later I will discuss the unique topology of organisms that allow them to escape this infinite regression.

Bateson propagated the idea that the usefulness of a map (*i.e.* a representation of reality) is not a matter of its literal truthfulness, but its having a structure analogous, for the purpose at hand, to the territory. That is, the usefulness of the sign is not its correspondence with the signified realm but its functional ability to do things. As I illustrated in the previous chapter, this ability to do things is the ability to mediate between two non-semiotic realms, be they the physical brain processes of two communicating agents, or DNA and proteins.

How can a sign functionally do things when it is a part of a closed semiotic system of signs and not a part of the realm it mediates? Let me explain this difficulty in semiotic terms.

Any system of signification, such as natural language, is a closed system in the sense that every legitimate operation within the system, on the units of the system, remains within its boundaries (Neuman, 2003b).

The systemic closure of semiotic systems explains why every utterance in a natural language is itself a part of the natural language, even if it is a meta-statement or a paradoxical statement that negates its own truth or existence. This systemic closure is self-evident because to violate it would ultimately destroy the boundary between the system and its corresponding realm. In other words, it would destroy the system's identity (Neuman, 2003b). Such a disastrous situation, in which the boundaries between the semiotic system and its corresponding realm blur or collapse, is evident in psychiatric cases when one mistakes the map for the territory (Bateson, 2000) or in children's fairytales when words materialize into concrete actions.

To exclude psychiatric cases and children's fairytales, *ipso facto* any semiotic system is clearly differentiated from the realm it signifies. When a sign turns into the thing it represents it looses its signifying power. On the other hand, signs must transcend the boundary of the system in which they are a part. Otherwise they would not be relevant to the other realm that they represent! How can one be both inside and outside the semiotic system at the same time? Elsewhere (Neuman, 2003b), I presented an answer to this question by pointing to the paradoxical nature of the sign as a boundary phenomenon that exists in between realms.

It might be intellectually intriguing to think of signification as a process that exists in between realms. Fortunately, physics provides us with a perfect analogue for understanding this in betweeness in terms of heat. Temperature and heat are two terms that are commonly confused by the non-expert. However, these terms are used to designate two different things. Heat is not a property of matter. Matter does not have heat but only kinetic molecular energy. Heat is energy in transfer and it exists only in between two systems and only as it flows from the hotter to the colder system. When one drinks a hot cup of coffee in a cold winter heat flows from the hot cup of coffee to the colder environment and never vice versa. A sign is similar to heat in several senses, but an important similarity is that both exist in between realms and both exist when information is lost. When information is lost and heat is released energy that could have done some work is lost. When we shift from a one-level differentiation to a two-level differentiation the variety of the first level is constrained and the combinatory potential of this level is restricted in favor of second-level order.

The existence of signs in between realms has a concrete manifestation in the genetic system. RNA's ability (*i.e.* mRNA) to "act as both genetic template [that points to the DNA] and biochemical catalyst [that points toward the proteins]" (Eddy, 1999) makes it a perfect candidate to serve as a genetic sign.

6. Why Do We Need a Meta-Language?

Let me turn again to the semiotic principles that underlie the need for a meta-language. In the process of transformation from a system such as the alphabet of the DNA to a semiotic system such as the RNA codons, something is lost and something is gained. On the one hand, a sign involves the loss of information in the sense that differentiated states collapse in favor of a more general and differentiated level abstraction (see the previous chapter). This loss of information is built into any process of computation unless it is designed as a reversible computing process in which theoretically no heat is released to the environment (Landauer and Bennett, 1985). On the other hand, a sign is highly informative and can be interpreted as referring to a particular class or object or can trigger a unique response. For example, the sign *tiger* preserves nothing about the color, the height, the gender, and many other distinguished features (*i.e.* information) of the particular carnivorous mammal to which it refers. However, the sign *tiger* can help the Indian farmer to run away when announced by his colleagues. No need for details. Just run!

A similar process is evident at the molecular level. Each codon of RNA is a sign that corresponds to three letters of the DNA alphabet. However, this sign is different from the DNA letters since RNA has a different base (*i.e.* U) and a hydroxyl group that gives the molecule *catalytic versatility that allows it to perform reactions that DNA is incapable of performing*. That is, the transformation from DNA to RNA is not a simple transcription such as the one of replacing the letters of a given alphabet with corresponding numbers. It is a transformation from a non-semiotic to a semiotic system in which certain information is lost in favor of signs—codons capable of performing reactions (*i.e.* being informative) in another realm, of triggering unique biological responses in a similar manner to signs in natural language. The different base and hydroxyl group is not a structural matter *per se* but a structural change that entails the potentiality of signification. In sum, the use of a sign necessarily involves the loss of information with regard to a particular instance of this representation, and on the other hand, the gain of information with regard to another realm. That is, the process of signification necessarily involves a shift to a higher level of abstraction. In this context, Bateson's insights are again indispensable for understanding the need for a meta-language.

7. Meta-Language is Inevitable

In one of his other seminal papers, "A Theory of Play and Fantasy", Bateson (2000) presents the idea that living communication systems operate at several levels of abstraction, and he differentiates between meta-linguistic levels of abstraction and metacommunicative levels of abstraction. The meta-linguistic levels of abstraction involve messages where *the subject is the language*. For example, the utterance "the word *cat* is not a cat" is a meta-linguistic message that says something about the meaning of the word *cat* and implies something about the status of signs in general. To review, a sign is never identified with the signified. People who have difficulties in moving between levels of abstraction and grasping meta-linguistic messages may confuse the map with the territory, sense and reference, or the sign with the signified. Those people might believe that the sign *cat* is really a cat or might eat the menu in a restaurant by mistaking it for the meal it signifies. The last example was used by Bateson to describe a schizophrenic patient who mistakes the sign (*i.e.* menu) for the signified (*i.e.* the food). Surprisingly, the link between pathology and the dysfunction of meta-language was recently discussed with regard to the genetic level. It was argued quite recently by Perkins *et al.* (2005) that:

> Altered regulatory control of the transcription or the translation of a gene may contribute to disease risk. (p. 2)

These researchers hypothesized that schizophrenia might be the result of this altered regulatory control as mediated by the ncRNAs. *In this case the schizophrenic patient mentioned in Bateson's example expresses the inability to use the meta-language on the behavioral level while Perkins and her colleagues identify the same difficulty at the genetic regulatory level!* As will be argued below, this convergence of ideas is not a coincidence but grasps a very profound truth of semiotic processes in general.

Metacommunicative messages are messages where the subject is the relationship between communicating agents. For example, after telling a joke that could have been interpreted as an insult, one may say, "I was just joking". For matter of convenience metacommunication and meta-language will be considered under the general title of *meta-linguistic processes.*

The importance of meta-linguistic processes was evident to Bateson when he observed young monkeys playing at the San Francisco Zoo. The interaction between the monkeys looked like a combat, even though it was not and the monkeys seemed to be well aware of it:

> It was evident, even to the human observer, that the sequence as a whole was not a combat, and evident to the human observer that to the participant monkeys this was "not a combat". Now this phenomenon, play, could only occur if the participant organisms were capable of some degree of metacommunication, *i.e.* of exchanging signals which would carry the message "this is play". (Bateson, 2000, p. 179, originally published in 1955)

Let me explain this argument. The monkeys played by pretending to fight. They exchanged messages of fighting although they were not. They exchanged signs that were untrue or not meant in the sense that they denote something that does not exist. After all, a signal of aggression during play does not really mean aggression. This playing activity is possible only by a supporting meta-linguistic frame that, on the one hand, allows the existence of those signs and, on the other hand, restricts their meaning. The monkey's signal says: This is a fight. But, the meta-linguistic level says the opposite: This is not really a fight! It is as if the monkeys had read Borges story and learned its lesson: a sign is never the signified!

Bateson made several important statements with regard to the meta-linguistic messages. He suggests that an important stage in the evolution of communication occurs "when the organism gradually ceases to respond quite 'automatically' to the mood-signs of another and becomes able to recognize the sign as a signal". That is, to recognize that the signals are only signals which can be "trusted, distrusted, falsified, denied, amplified, corrected, and so fourth" (Bateson, 2000, p. 178).

This meta-lingusitic ability, which Bateson counter intuitively conceived as *preceding* the denotative power of signs, establishes a paradoxical frame in which map-territory/sign-signified relations are both *equated and discriminated* within the same activity. This is the reason why meta-language necessarily accompanies language. A sign holds a paradoxical nature in between realms but knowing how to live with the paradox is not a simple matter and meta-language is needed to help us.

Let me further elaborate these ideas. There are two complementary aspects of paradoxical activity in which the sign signifies something (equates, *e.g.* this is a fight) and at the same time denies this signification (discriminates, *e.g.* this is *not* a fight). First, the meta-linguistic level always qualifies the referential power of the sign (restricts its reality, according to Bateson) and therefore opens the way for a variety of interpretations (If it is not a fight, what is it? Can it be something else? Maybe a game?). For example, a codon usually corresponds to a specific amino acid. However, the correspondence between the codon-sign and its corresponding amino acid realm is not a simple one-to-one correspondence. Variations on codon correspondence, although statistically rare, do exist (Kanehisa, 2000). As I mentioned earlier, CUU usually correspond to Leu. However in the yeast's mitochondrial code it corresponds to Thr. In other words, a cigar is sometimes just a cigar but as a sign it has the potential of corresponding to many other things, such as signifying the human phallus. This Polysemy of a sign system is a property that endows the system with enormous flexibility, and results from detaching the sign from its concrete embodiment (the sign is not the signified) and increasing its entropy to maximum (My God! If it is not the signified, it can be anything!). Later I will discuss this property under the title of *arbitrariness*. Indeed, by detaching the sign *cat* from its concrete perceptual instances, certain information was lost on one level of analysis but was gained on a higher level of analysis. The sign *cat* may be used to denote a Jazz player or any meaning speakers of slang may invent. In one of the final chapters of the book, I will discuss the paradoxical nature of the sign in terms of a superposition. This discussion will allow us to add intellectual depth to our semiotic analysis.

While the meta-linguistic level let the monkeys in the above example understand that "This is not a fight", the signs that are exchanged by the monkey signify: "This is a fight". Therefore, taken as an isolated object, the sign presents the other extreme position of signification in which the sign points directly and might even be identified with the signified (This is *really* a fight!). This position is necessary but, taken in isolation, is unbearable from a semiotic point of view, since it violates the rule that signifiers are always pointing at something else as well as the idea that the sign is loosely (or arbitrarily) associated with the signified. This *frustration* is constructive the same as the *frustration* of proteins is constructive in directing their folding (Shea *et al.*, 2000). Let me explain the constructiveness of this frustration.

From a general semiotic point of view, the two extreme positions (sign = signified and sign ≠ signified) are by definition impossible in isolation, but complementarily necessary. *Therefore, through the meta-language the sign is established as a unique entity that exists in between the two realms*. Metaphorically speaking, through the meta-linguistic level the sign is in a state of *superposition* that endows it with the power to transform the closed semiotic system into a corresponding non-semiotic realm without being an integral part of any system. Again, the idea of a sign being informative while existing in between in a superposition may sound awkward. However, the whole idea of quantum computing is based on particles that in contrast to the binary nature of classical information (*i.e.* 0 or 1) may exist both at 0 and 1 at the same time.

8. The Importance of In Between

The linguistic realm is not the only case for illustrating the importance of existence in between. In mathematics the introduction of imaginary numbers represents the same form of existing in between. What is an imaginary number? An equation such as $x^2 = -1$ does not have a solution in the realm of rational numbers since the square of any real number is never negative. The solution offered by mathematicians is the introduction of the symbol i by defining i^2 as equal to -1. That is, i is the square root of minus one. This imaginary number, i, is imaginary not only in the sense that it does not correspond to countable objects in the natural realm but in the sense that it signifies an impossible object.

The mathematician George Spencer-Brown (1994) made an astounding observation about the nature of the imaginary number. Spencer-Brown (1994) suggests that the expression $x^2 = -1$ can be written as $x = -1/x$ and points out that this is a *self-referential* expression like the paradoxical statements in logic we are familiar with:

> The root-value of x that we seek must be put back into the expression from which we seek it. (p. xv)

If we assume a world of binary information values, then x should be either $+1$ or -1. No other meaningful alternative exists in this classical view of information. If $x = +1$ then $+1 = -1/+1 = -1$ is clearly paradoxical. If $x = -1$ then $-1/-1 = +1$ is equally paradoxical. The imaginary number introduces time to our system as was realized by Spencer-Brown and from a different perspective by the psychoanalyst Matte-Blanco (1988). I will discuss this idea in the concluding chapter by drawing on Deleuze's idea of repetition (one-level differentiation) as a paradox echoing and returning on itself, and constituting the sign system.

The power of the sign is like the power of the imaginary number. In both cases we have something that does not really belongs to a binary realm. It is a paradoxical entity that is constituted by a recursive function, exists in between, and has an indispensable value for various operations. However, as a paradoxical entity that exists in between there is always a need for a meta-frame to support and constitute it in various ways.

In sum, as something that exists in between realms (*e.g.* DNA and proteins) the biological sign of RNAs is of a paradoxical nature that can be regulated outside of the system through a meta-level in order to endow it with the power to do things. This semiotic mechanism holds both for the biological and the linguistic realm. Understanding this logic and using it in the study of genetic phenomena may provide biologists with a powerful tool to think with while researching "unthinkable" research.

9. Conclusions

What are the general conclusions we may draw from the above discussion? The first conclusion concerns the metaphorical nature of the linguistic metaphor in genetics. I critically examined this metaphor but I do not dismiss its value. My argument is that there are general semiotic principles that are evident in various biological and linguistic systems. Biological systems do not have language in the same way that human beings have language. However, both systems obey general principles like the requirement that any language will be accompanied by a meta-language. The implication of this insight for reductionism is clear. Biological systems cannot be reduced to the genetic language because this language is accompanied by a meta-language that exists at a higher logical level of analysis.

The second conclusion concerns a new perspective for studying ncRNAs. As time passes more information is gathered about the various functions of the ncRNAs. However, without an appropriate theoretical framework the data collected is to a large extent meaningless. The idea presented in this chapter is a possible theoretical framework for examining the ncRNAs. I do not pretend to present the only theoretical framework or the ultimate theoretical framework for studying the ncRNAs but just one perspective that may be theoretically beneficial.

What are the benefits of using this perspective? Considering ncRNAs as a meta-language may direct us to study the details of metacommunication in both the linguistic and the biological realm. For example, we may better understand the logic behind poorly understood genetic phenomena such as DNA methylation, which is directed by RNA (Chan *et al.*, 2004; Mattick, 2001; Wassenegger, 2000). Methylation is the addition of a methyl group to a cytosine residue of DNA to convert it to 5-methylcytosine. This process involves the operation of an enzyme that attaches a methyl group to carbon 5 and alters its properties. Methylation has an important role in the development of eukaryotic cells and in mammalian epigenesis (Jones and Takai, 2001; Reik *et al.*, 2001). For example, it was argued that cytosine DNA methylation silences harmful DNAs such as retroviruses. In other words, the transcription from DNA to RNA is mediated by the methylation process which is regulated by ncRNAs. This process will be discussed later when I discuss silence from a pragmatic perspective. In this context the role of meta-language will be more comprehensible.

By approaching DNA methylation through the lenses of meta-language we may uncover similarities, discrepancies, and unclear mechanisms in both the biological and the linguistic realm. One may hardly find in the genetic research an explicit and elaborated form of this meta-linguistic perspective. In this context, the idea of ncRNA as a meta-language is at least justified as food for thought. I must admit that personally I will be satisfied to supply this food. The next chapters aim to move us forward in examining a semiotic alternative to biological reductionism. This time I take immunology as my field. Before getting into immunology let us take a break to converse with the cat.

Cat-logue 2

Dr. N: Okay, my dear cat, let's summarize the previous part of the book.

Bamba: Well if I correctly understood your argument we might have a certain problem.

Dr. N: Why?

Bamba: You say a process of computation is a process in which information is lost.

Dr. N: And?

Bamba: Reading comprehension is a process of computation in which structures are produced from processing textual information. Therefore reading comprehension involves the loss of information. If I follow your logic then I am currently in a state in which after reading the previous chapters, I perfectly understand your thesis but cannot recall what it is about. I simply lost all the information!

Dr. N: Very funny. I almost forgot your unique sense of humor …

Bamba: … but as you emphasized again and again, life means paradox, and we should learn to live with paradoxes rather than to solve them.

Dr. N: Very, very nice! You seem to be a much more intelligent creature than smart Hans, the horse that excited the imagination of Europe by solving simple arithmetic problems.

Bamba: Don't even try to compare me with that horrible creature. Horses, like human beings, are social animals, and living in a herd does not provide the perfect conditions for creative and autonomous thinking. In fact, I don't have to tell you this. After all, you are a part of the university herd and …

Dr. N: Enough! The faculty's Dean might read this book and the loss of my position would have painful consequences for the quality of your diet.

Bamba: As a pragmatist, I perfectly understand this position. Please send my apology to the Dean and let's summarize the previous chapters.

Dr. N: This is a constructive move!

Bamba: Ok. The first thesis you introduced is that organisms are irreducible because they are "machines" that compute themselves and during the process of computation information is necessarily lost.

Dr. N: Correct.

Bamba: But why do you call organisms "machines"? Are you a mechanist? God forbid! Do you believe that organisms are like toy machines?

Dr. N: No. A machine is just an abstract idea suggesting that the system under inquiry is something that does some work.

Bamba: And what is work?

Dr. N: *Work* is also an abstract idea describing the way in which energy is utilized …

Bamba: … and *energy* is of course another abstract concept that is no less comprehensible than the other concepts you introduced.

Dr. N: I'm afraid that there is no way in which we can truly escape from this labyrinth. However, there is magic in this activity in the sense that it brings you to some interesting places. Think about the idea of the organism as a computing machine. This idea forces us to examine the meaning of producing a certain output from a certain input. If the genome is the input of the system then according to the physics of computation some information will be lost during the computation process. This is an interesting problem that emerges from our conceptualization.

Bamba: So what are you doing? Creating problems rather than providing solutions? Is this is the work of a scientist?

Dr. N: Definitely! Unless you are a naïve realist who believes that scientists are people who tear the curtain masking reality or post-modernists who believe that scientific theories are just discursive structures. Scientists create problems. They imagine them and then see what work can be done with these imaginings, their implications, and their solutions. For example, the idea that the organism is computed using the genome as a source of information, a set of differentiated entities that take a part in the biological construction process, explains to us why organisms are irreducible. This explanation is the work done by our imagination.

Bamba: So let me ground you with a specific question: Are we stardust?

Dr. N: The building blocks of organisms may be an evidence of our galactic history. However, stardust does not simply turn into a living creature.

Bamba: Sure. We have to assume random …

Dr. N: Ahh! How many times did I tell you that God does not play with dice, and that our natural history was not born in Las Vegas? The idea of randomness is just a way of formalizing our ignorance; it is not an explanation. Let's stick to our idea of organisms as machines that compute themselves without getting into speculations about the origin of species.

Bamba: Help! Can someone help me; there is a creationist in this room!

Dr. N: Hush! I am not a creationist but I resist dogmatic thinking. Read Darwin. In the *Origin of Species* he writes (Darwin, 1958, p. 131):

> I HAVE HITHERTO sometimes spoken as if the variations—so common and multiform with organic beings under domestication, and in a lesser degree with those under nature—were due to chance. This. Of course, is a wholly incorrect expression, but it serves to acknowledge plainly our ignorance of the cause of each particular variation.

In fact my criticism of neo-Darwinism is that it has not biologized biology enough. By the way, I consider human being to be the best evidence for supporting the theory of evolution. It is impossible that intelligent design created such a destructive creature.

Bamba: Ohh! I feel relief. I was afraid that you were a creationist and that we were moving to Alabama.

Dr. N: Alabama?! Don't worry. We are staying in Israel. The heart of the storm is the safest place. So let's move to the second point concerning the relation between language and meta-language.

Bamba: This was an interesting chapter. I never imagined that junk DNA is not so much junk. Do you think that there are other forms of junk that are not junk? How about junk food? Maybe McDonald's hamburgers are, after all, an important component in our nutrition?

Dr. N: I'm afraid you are missing the point. The idea that meta-language is a complementary aspect of language shows that our unidirectional conception of biological processes is wrong. Organisms are machines that compute themselves by maintaining an orchestrated balance between language and meta-language. In this sense, organisms cannot be studied top-down or bottom-up. They are recursive-hierarchical structures that should be studied as self-generating and self-constituting wholes.

Bamba: The logic of in between?

Dr. N: Indeed, and the immune system is the next case I'm using to illustrate this logic. In the following chapters, I present a brief introduction to immunology and two chapters that consider unsolved problems in immunology from a meaning-making perspective. In both cases, I will show that it is possible to think about the immune system as a meaning-making system and that this way of conceptualizing the immune system may be theoretically interesting.

Bamba: I'm totally confused. I assume that meaning, whatever it is, is the result of a computation process and, as such, involves the loss of information in favor of abstraction. Metaphorically speaking, making sense is transcending the trees to see the forest, but if I am a cat that lives in the forest then … OHHH!!! Can't you develop a more user-friendly thesis?

Dr. N: Unless you prefer to be the cat of Prof. Dawkins, I suggest that you will keep up with the obscurity of my ideas.

Bamba: As a selfish cat that is ruled by his selfish genes and wishes to be nurtured by a selfish master, who is a university professor with a selfish ego, I have no other choice than to keep reading the next chapters.

Dr. N: You see how simple recursive logic can be!

Chapter 6

Immunology: From Soldiers to Housewives

The immune system is our second case for examining reductionism. What is the immune system? A very simple, and necessarily inaccurate, answer is that the immune system is the organism defense system. As defined by Marchalonis and Schluter (1990) with regard to the immune response:

> We consider the immune response to be a subset of defense mechanisms that expresses certain defined properties. (p. 758)

Why is this general definition inaccurate? The reason is simple. It is paradoxically too broad and too narrow at the same time, like any other general definition. This is the reason why definitions are of help to those who already know the phenomenon and just want to articulate their knowledge in a concise way. To those who are not familiar with the phenomenon a definition may not be a good starting point. Let me therefore begin with the association the immune system usually evokes among naïve readers. The immune system is visualized in our mind as a team of white blood cells (*i.e.* leukocytes) that fight vicious bacteria and viruses that try to invade our body. As a child I learned about the immune system from a picture book that portrayed the white blood cells as marine soldiers who patrol the blood stream in speedboats and exterminate vicious invaders.

This picture of the immune system is not wrong but just oversimplified. The immune system does have agents that fight the invasion of pathogens and the immune system is involved in defense activities. However, the immune system is also considered as a diffuse sensory system and it is also involved in less heroic activities but no less important activities such as housekeeping (Cohen, 2000b). It is a system of well-defined biological agents that cooperate in defending and maintaining the organism. It is an amazing system and I hope to communicate my enthusiasm through this part of the book.

1. The Innate and the Adaptive

Introductory books like to present the immune system by differentiating between two types of immunity: innate immunity and specific (adaptive) immunity. Although in practice the two systems are intermingled, for instructional purposes only, I will stick to the oversimplified presentation of the textbooks.

Innate immunity is composed of three basic lines of barriers. The first line is composed of physical and chemical barriers. For example, our skin is a physical barrier that protects us from the invasion of outside intruders. It is composed out of an external and insensitive layer (epidermis) and a deeper and sensitive layer known as the dermis. We realize the important defensive role of the skin in cases of severe burns where the body becomes exposed and the physician must cope with the threat of severe infections caused by the intrusion of pathogenic agents that cross the first line of defense.

The second line of defense is composed of proteins that concentrate around different entrances to the body. For example, the saliva in our mouth functions to help us to digest the food we eat. However, the saliva also contains lysozyme enzymes that fight pathogens by breaking down their membrane. I assume most people have never imagined how deadly saliva can be. Observing wounded mammals that lick their wounds it is clear however that the saliva has a defensive role. Moreover, it was found that even insects such as moths produce lysozyme as a response to bacteria (Marchalonis and Schluter, 1990). This finding may serve as an indication to the common evolutionary origin of this specific defense mechanism. Indeed, it was found that the lysozyme molecules of humans, chicken, and the moth *Cecropia* have a homologous (*i.e.* a corresponding) genetic structure. Personally this finding warms my heart because I always considered Chimps to be my kin and suddenly my family was broadened to include the moth Cecropia!

The third line of defense involves phagocytic cells (*e.g.* macrophages, natural killers, neutrophils) that have great fun in swallowing vicious invaders. Phagocytosis is the same phenomenon as the engulfment of detritus or bacteria by free-living amoeba. It is a universal mechanism in the animal kingdom (Bayne, 1990) and the predominant defense mechanism in many of the most primitive animals. Eating your opponent—Cannibalism—seems to have roots in phagocytosis.

At first, the idea of phagocytosis seems quite simple: Foreign particles that enter the organism are swallowed by certain cells. On second thought the process seems much more complex. How does a phagocytic cell "know" which cells it should swallow? The answer to this question encapsulates a much wider quandary that bothers immunologists to this day. How does the immune system "know" to differentiate between self and non-self, between constituents of the host and foreign invaders? This issue will occupy us at length in the next two chapters. Meanwhile, let me say a word on phagocytosis and non-self-recognition. Recognition of foreign or damaged cell tissues can be done *directly* by recognition molecules, such as the immunoglobulin antibody, which will be discussed in the following text. Recognition can be done indirectly through agents (*i.e. opsonins*) that bind to the damaged tissue or the foreign invader and mark it as a non-self.

These lines of defense compose the primitive part of our immune system and we share them with lower-order organisms such as the sponge, which is the simplest multi-cellular organism (my apologies to SpongeBob, the favorite cartoon hero of my kids). Innate immunity is characterized by a relatively low specificity for microbes, limited diversity, stereotypic form of activity with limited specialization, and with no memory at all. In other words, the components of innate immunity are quite limited, have not specialized in identifying specific pathogens, and do not remember their enemies. It is a very general and simple defense system.

In contrast with innate immunity, specific or adaptive immunity, which is evident in higher-order organisms, has different characteristics. Adaptive immunity is highly specific in its sensitivity to distinct molecules, it has a large diversity of agents that are highly specialized, and memory (which is materialized through memory cells) that allow it to remember and respond more effectively to repeated exposure to the same pathogens. Let me dwell a little bit on these characteristics.

Adaptive immunity is specific in the sense that immune responses are specific to distinct antigens (agents that trigger the immune response). Antigens have structural markers that are identified by the immune agents. These markers are called *determinants* or *epitopes*, and the specificity of the immune response is made possible by the ability of B and T immune cells to distinguish between these antigens. The diversity characteristic concerns the enormous diversity of lymphocytes (*i.e.* white blood cells) that can distinguish between different antigens. Our immune system can potentially differentiate between approximately 10^9 different epitopes! I doubt whether there is a more sensitive recognition system in the universe.

The next characteristic is memory. Like us, the immune system keeps traces of the past. Specifically, memory cells exist for long periods and are ready to respond rapidly to vicious antigens they have already encountered in the past. In one of the final chapters I return to the issue of immune memory and elaborate on it with regard to other important concepts of meaning making including "context".

The last characteristic is specialization. The immune system not only recognizes different antigens through diverse agents but responds in a distinct and specialized way to different antigens.

It is interesting that diversity of components, specificity of recognition, specialization and memory, are not only characteristics of adaptive immunity but of other adaptive systems as well. Think about human intelligence. Intelligence is a vague and loaded concept with too many disturbing associations and connotations. However, intelligence in the most general sense can be considered as the ability to successfully solve a variety of problems. Although we may try to solve different problems with the same general strategy, this approach is of limited success. Please remember this important point since it will occupy us in one of the next chapters.

I painfully realized the limits of a general recipe for problem solving when I was in junior high school. As a junior high school student, I hated math lessons, specifically algebra word problems, and desperately searched for a recipe that will help me to avoid the burden of solving my math homework. While I hated math, I loved to read and one day a "brilliant" solution popped into my mind! I read a famous treatise by the French philosopher Rene Descartes. This treatise—Mediations—offers the reader with a way to establish his knowledge on solid grounds through general principles of thought. For example, one of the recommendations offered by Descartes is to divide any problem into the smallest elements you can. Today, we call this tactic *analysis* and it is a cornerstone of the reductionist approach. At that time the reductionist approach seemed promising. I was excited. No more boring algebra word problems. I had a method, a general key, to all the problems that I would encounter from then on. It is unnecessary to say that I was a total failure at the following math exam in which I attempted to apply Descartes' mediations. However, I learned a lesson on the shortcomings of overly general, problem-solving strategies.

Recognizing the specificity of a problem is important for offering a specific response through a variety of means. This is a general characteristic of intelligence, both human and immune. This is also the difference between a witch doctor and a modern physician. Anyone who has had experience with physicians knows how delicate this difference may be. However, for the sake of didactics we can say that the witch doctor has a relatively limited capacity for diagnosis (*e.g.* "Your chakras are blocked"), treatment (*e.g.* "Let me open your charkas"), and no memory at all. In this case, the lack of memory means the lack of scientific and critical recordings of previous cases. After all, if all the medical problems can be exhausted by blocked chakras why should the witch doctor keep records? Why should he or she care about accommodating knowledge? The witch doctor is like the innate immune system. In contrast, a modern physician uses records of medical knowledge held by the thought collective (*i.e.* previous encounters with the disease and the lessons learned by members of the medical community). Therefore, he or she has a great variety of known symptoms from which to make a specific diagnosis and to provide a specific treatment.

The picture I have just portrayed might be a little bit misleading. It is not the case that in higher-order organism only specific immunity exists. Adaptive immunity is a system that operates in cooperation with the innate immune system and does not cancel it. It is another layer of complexity added onto immunity and not a substitute.

2. The Agents of the Specific Immune Response

What are the agents of the specific immune response? The first agents are the lymphocytes. Lymphocytes are the only cells that can specifically recognize antigens. There are two major classes of lymphocytes: B lymphocytes and T lymphocytes. B cells are produced in the bone marrow and T cells are produced in the bone marrow but migrate to and mature in the *Thymus* a lymphoid organ, which is anterior to our heart. This maturation process is interesting since it shows that recognition, the B cells' primary task, is something that must be learned. Indeed, immune recognition, rather than a simple process of pattern recognition, seems to be a process that involves learning, and learning is the process through which sense making is developed.

B cells produce a specific molecule—antibody—that plays a crucial role in the identification of the antigen. The major function of T cells is in the regulation of immune response and as effector cells for the extermination of intracellular microbes. In addition, there are other players in the game. Macrophages (*i.e.* mature monocytes that settle in the tissues after migrating from the bone marrow) play a role in the innate and specific immunity. For example, they are involved in the swallowing (phagocytosis) of foreign particles such as microbes or even injured or dead self tissues. They also produce *cytokines* (*i.e.* proteins that serve as intercellular mediators) that recruit other cells involved in inflammation such as neutrophils. Macrophages also play a role in specific immunity. For example, they display antigens on their surface in a form that can be recognized by the T cells. In professional terms they are *Antigen Presenting Cells* (APC). This is an incredible activity since it clearly involves cooperation and communication between cells. One cell presents something (*i.e.* antigen) to another cell in a way that the addressee of the message will understand the message.

There are other cells that are involved in immune activity but they are not directly relevant to our inquiry. The interested reader is invited to read Cohen's (2000a) book on immunology for further details.

3. Immune Recognition

Immune recognition is a fascinating issue and an issue where the limits of classical reductionism can be illustrated. In this section an introduction to immune recognition will be provided and the associated difficulties will be presented.

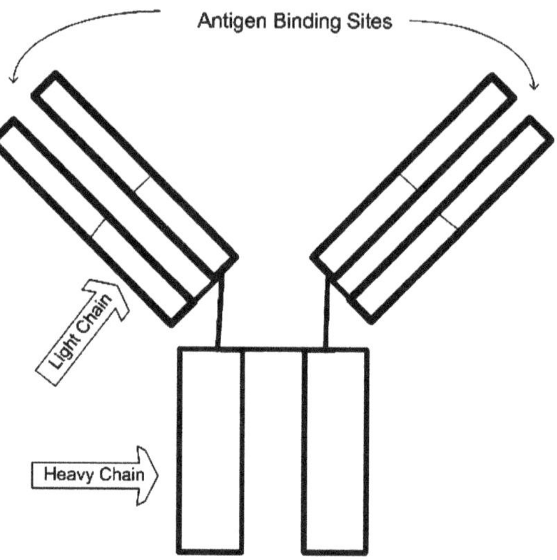

Fig. 6.1 A schematic representation of the antibody.

As we previously learned, the B cells play a crucial role in immune recognition by producing antibodies. The structure of the antibody molecule (*i.e.* immunoglobulin molecule Ig) is interesting and genetically unique. Therefore, I will present it first. A schematic structure of the antibody appears in Fig. 6.1.

It can be seen that all antibodies have a core structure of two identical light chains and two identical heavy chains that are attached to each other by disulfide bonds. The light chains are divided into two classes—*isotype*—with no functional distinction between them. The heavy chains are divided into five different isotypes. These isotypes are: IgG, IgM, IgA, IgD, and IgE. For the sake of brevity my discussion concerns immunoglobulins from the IgG class, which is the major type of antibody in a normal human being.

The light and the heavy chains are divided into constant and variable regions. The constant region is responsible for initiating effector functions and the variable regions are responsible for the recognition of the antigen. The enormous potential diversity of the antibodies is structurally expressed in three short segments of the heavy and light chains. The highly divergent regions in the variable area are called hypervariable regions or complementarily determining regions (CDR1–CDR3). These regions constitute the *binding site* for the *antigen* in the 3-D folded antibody.

The existence of antibody diversity presented researchers with a quandary. As we previously learned, the structure of protein molecules, such as the antibody, is determined by the sequence of DNA. However, there are *two fascinating exceptions*: the receptors of the B and the T cells. Why they are exceptions? Because:

> You do not inherit the DNA genes that encode your antigen receptors; you manufacture your own receptors genes epigenetically from genetic raw materials. (Cohen, 2000a, p. 144)

In other words:

> B cells rearrange the genes that code for their antibody protein, so that each cell makes a unique antibody. (Branden and Tooze, 1999, p. 300)

Let me explain this fascinating mechanism for generating antibody diversity. The source of the antibody's diversity is a small segment of DNA. There are three gene pools that encode for the variable and constant regions. One pool codes for the heavy chain and the other pools for the two isotypes of the light chains. There are three gene segments that encode for the heavy chain: V (1000 different segments that code for the first 90 residues), D (10 different segments that code the hypervariable region CDR3), and J (four different segments that code for the remaining 15 residues of the variable domain).

The DNA found in the variable region of a new B cell is constructed by a "random joining of one of each of these segments into a single continuous exon" (Branden and Tooze, 1999, p. 302). This combinatorial joining creates a new exon that joins one of the C segments that encodes for the constant regions of the heavy chains. Later, variable regions will be exchanged among heavy chains, a class-switching process that will add complexity to our observation. A similar process is evident in the light chain but with no D segments and therefore the diversity of the light chain will result from the joining of the V and J segments. This is not the end of the story. Shuffling the genetic cards of the antibody is a bottom-up process. However, mutations can be introduced into the exons of the variable domains through *somatic hypermutations*—alteration of a germline immunoglobulin sequence by introduction of nucleotide changes during the lifetime of a B cell (Wagner and Neuberger, 1996). Complexity is not an indication of the limits of our knowledge; it is built into nature.

The T cells have a different recognition mechanism. T cells recognize antigens only when antigens are presented with MHC protein molecules. These proteins are a part of the major histocompatability complex. This complex is a region of chromosome six (in human beings) that was identified as controlling transplantation rejection. It is a unique genetic marker of each of us. Each of us has a unique genetic marker of his/her self that prevents transplantation from another person. The T cell receptor is unique in the sense that it can specifically bind to the MHC, which is presented on the surface of a potential antibody.

We may turn now to the actual activity of recognizing the antigen. Basically an antigen is any molecule that can bind to the antibody. If these molecules result in an immune response we call them *immunogenes*. The recognition is not an abstract process taken place in the "mind" of the molecule. The recognition involves the binding of the antigen to the antibody through non-covalent forces. In other words, the binding is reversible. But how does an antibody know to whom to bind? This question is discussed in the next chapter where the enigma of immune specificity is addressed.

Chapter 7

A Point for Thought: Immune Specificity and Brancusi's Kiss

Summary

Immune specificity is usually described in terms of the lock-and-key metaphor. However, this metaphor is to a certain extent misleading and does not convey the complexity underlying immune specificity. The failure of the lock-and-key metaphor makes it difficult to understand immune specificity and recognition. This is the reason why immune specificity has been described as the *specificity enigma.* In this chapter, I point to three important differences between biological specificity and the mechanical specificity that underlies the lock-and-key metaphor. I further suggest an alternative lens through which immune specificity can be considered.

1. On Miraculous Drugs and Biological Specificity

Several years ago a family member shared with me a medical problem her son had. After the family physicians failed to solve the problem, she started looking for solutions in alternative medicine. Enthusiastically she told me that she might have found a solution to the problem: a new promising "natural herbal product" which is "good for everything". My immediate response was, "If it is good for everything then it is good for nothing". "Why?" She asked me. My answer pointed at the specificity of molecular mechanisms as a scientific fact. However, I could not simply dismiss her naïve question. After all, biological specificity is usually described in terms of the *lock-and-key metaphor*. If this is the leading metaphor why should we dismiss the possibility that there is a general *master key* that can open all the locks, a miraculous drug that can address all medical problems? Unfortunately, and for good reasons, such a miraculous drug does not exist. This incidence drove me to a reflective examination of the lock-and-key metaphor and brought me to some interesting conclusions concerning immune specificity. Let me open my discussion by first introducing the metaphor.

The lock-and-key metaphor was introduced in 1894 by the Dutch biologist Emil Fisher who proposed that enzyme and substrate fit together like a lock and key. As argued by Clardy (1999):

> No analogy has so profoundly influenced our thinking about the joining of biological molecules as Emil Fischer's lock and key. (p. 1826)

However, in certain cases there are serious problems with this metaphor, and biologists should re-consider their use of the metaphor and seek alternatives. In this context, I would like to address the challenge of differentiating between *mechanical* and *biological* specificity in the context of immune recognition and point at meaning making as an alternative approach to immune specificity. This meaning-making perspective will be presented in a simplified manner to which depth and complexity will be added in the next chapters.

2. Specificity in Immune Recognition

Let me elaborate on the notion of immune recognition. As previously described, a major and crucial phase of immune recognition involves the binding of an antibody (*i.e.* an immunoglobulin) to an antigen (Abbas *et al.*, 2000; Cohen, 2000a). Recall, antibodies are protein molecules that function as the receptors for the B lymphocytes and bind to a specific antigen through non-covalent forces. Antibodies have a similar core structure with two identical light chains and two identical heavy chains. The light and the heavy chains are composed of a series of repeating units, and each individual is populated by an enormous approximate potential number of 10^9 different antibody molecules with a unique sequence of amino acids in their combining sites. There are variable, hypervariable, and constant regions in the antibody. Three hypervariable regions of the light chain and three hypervariable regions of the heavy chain fold to constitute the antigen-binding site. Ligand-receptor binding is conducted when the antibody binds to the epitope, which is the combining site of the antigen.

Ligand-receptor binding is commonly described in immunology using to the lock-and-key metaphor (Abbas *et al.*, 2000). However, Cohen (2000b) points to the difficulties of this mechanical metaphor by discussing four characteristics of immune recognition:

1. Degeneracy
2. Pleiotropia
3. Redundancy
4. Randomness.

Degeneracy refers to the "capacity of any single antigen receptor to bind and to respond to (recognize) many different ligands" (Cohen, 2000a, p. 138). The degeneracy of antibodies clearly presents a theoretical problem for those who use the lock-and-key metaphor (Cohen *et al.*, 2004). A one-to-one unique correspondence between an antibody (lock) and its corresponding epitope (the key) does not exist. In living systems monogamy is excluded, at least at the molecular level.

Pleiotropia is used to denote the capacity of an agent to produce many diverse effects and *redundancy* to denote an effect that is produced by several diverse agents. Pleiotropia and redundancy add another layer of complication to the way immune specificity is achieved. If the same cell can bind, interact with, and effect several other cells, and, if the same agent can be affected by several other agents, then specificity cannot be attributed to the *structure* of the molecule *per se*. *Randomness* concerns the epigenetic and somatic construction of the binding site from the three families of gene segments (V, D, and J). If the construction of the binding site is done through random combination of basic genetic units then the ability of the lymphocytes to recognize antigens is potentially unlimited. It is important to remember that the antigen receptors of the B and T cells are manufactured epigenetically from genetic raw materials (Abbas *et al.*, 2000; Cohen, 2000a). Therefore it can be argued that immunological specificity is not completely inherited but actively created. In this sense, and to use poetic language, when studying immune specificity we should shift our focus from mechanics to poiesis, the Greek term for creation.

The conclusion we may draw from the above analysis is that although specificity is evident in immune recognition, the lock-and-key metaphor (*i.e.* mechanical specificity) is inappropriate for describing it. This is the reason why Cohen decided to describe the specificity of immune recognition as the specificity enigma (Cohen, 2000a) and to consider immune specificity as an *emergent property of a complex immune system.*

If immune specificity is actively created rather than mechanically determined what is the alternative to the lock-and-key metaphor? The following sections accept the idea that the immune system is a complex cognitive system as argued by Cohen, and suggest that within this framework immune specificity should be discussed from a meaning-making perspective.

3. Specificity as Meaning Making

To critically examine the relevance of the lock-and-key metaphor and to offer an alterative, we should be aware of several crucial differences between mechanical and biological specificity. There are many differences between mechanical and biological specificity. In this chapter, I would like to discuss three major differences under the titles: recursive-hierarchy, synsymmetry, and hypothetical inference.

3.1. Recursive-Hierarchy

The first crucial difference between mechanical and biological specificity is that *mechanical specificity is achieved without crossing any scales of* organization. The key and the lock are two entities that exist as differentiated objects on the same level of organization. When there is a match between the geometrical properties of the key and the lock, the lock is opened. In contrast, immune specificity *crosses scales of analysis* through bottom-up and top-down signaling processes that are orchestrated by feedback loops. This dynamics was identified long ago in the pioneering work of Gregory Bateson and discussed under the title of recursive-hierarchy (Harries-Jones, 1995). It was also identified in different fields under different titles. For example, in hermeneutics the concept of hermeneutical circularity clearly resembles recursive-hierarchy. In biology, Conrad (1996) coined the term "percolation network" to describe the dynamic in which macroscopic inputs percolate downward to influence microscopic states and the way microscopic states percolate upward to influence macroscopic states. In one of the following chapters we will delve deeply into this concept and show how important it is for understanding living systems.

Cross-scale interactions within a functional whole seem to be a constituting principle of biological and cognitive systems alike. In this context, immune recognition is not an exception and recursive-hierarchy may be a powerful concept for explaining other biological processes. Immune recognition involves cross-scale interactions: from the network of non-covalent forces that bind the atoms of the proteins that constitute the binding site to the interactions in which the recognition takes place. Therefore, in order to understand immune specificity we should study cross-scale interactions and the boundary conditions that control these interactions. However, currently we lack the appropriate metaphors or conceptualization scheme for guiding this inquiry. As previously suggested, the boundary conditions of living systems are constituted through semiotic activity. Following this line of reasoning it is trivial to study immune specificity from a semiotic perspective. In this context, the linguistic metaphor naturally pops-up into our discourse again.

Cross-scale interactions clearly resemble the process of meaning making in text comprehension where microscopic particles of the text (*i.e.* words) influence the macroscopic text as a whole, and the macroscopic or whole text provides the appropriate context for understanding the meaning of single words. Can meaning making be an alternative lens for considering immune specificity?

Broadly speaking, *meaning* will be defined as the effect of signs-mediated interaction, and *meaning making* as the process that yields this effect. For example, the linguistic sign *cat* by itself is devoid of specific meaning. It is indeterminate because it can be used once to describe a certain animal and once to describe the nickname the local press gave to a skillful burglar. The sign *cat* becomes meaningful only through cross-scale interactions with other units and levels (*e.g.* sentences, paragraphs) that comprise the whole text. In a similar way, the meaning of an antigen is not solely determined by the structural properties of the suspicious molecule but through a complex process in which many immune agents are involved across scales of analysis (Cohen, 2000a).

In contrast to living systems, mechanical systems are not involved in meaning making. Metaphorically speaking, the meaning of a key is predetermined by its geometrical properties and their mirror image in the lock. No mediation is evident. No interaction is evident. No context is evident. It is a simple mechanical encounter. In this sense, one may predict the response of the lock to a given key (opened vs. closed) based on a simple structural analysis of the two entities prior to any interaction between the two. Along the same line, immune specificity cannot simply be predicted from a structural analysis of the units involved in the binding. It is an emergent property that results from cross-scale and semiotically mediated interactions, similar to those that characterize a text and the interactions between a reader and a text (*i.e.* text comprehension). The idea that in mechanical specificity the meaning of the key is predetermined by its geometrical properties brings us to the next issue of *synsymmetry*.

3.2. Synsymmetry

The second difference between mechanical and biological specificity concerns the issue of *symmetry*. To understand this issue one should be familiar with the way a key opens a lock. For a short introduction to keys and locks refer *The MIT Guide to Lock Picking* (1991).

A mechanical key is inserted into the *keyway* of the *plug*. *Wards* (the protrusions on the side of the keyway) restrict the set of keys that can get into the plug. The plug is a cylinder that can rotate when the proper key is fully inserted. The non-rotating part of the key is the *hull*. The proper key lifts each pin pair until the gap between the *key pin* and the *driver pin* reaches a *sheer line*. When all the pins are in this position the plug can rotate and the lock can be opened. Figure 7.1 is a schematic representation of a lock-and-key matching (*The MIT Guide to Lock Picking*, 1991):

As can be seen, only the appropriate one-to-one correspondence between the key and the driver pin can lead to the opening of the lock. The alleged

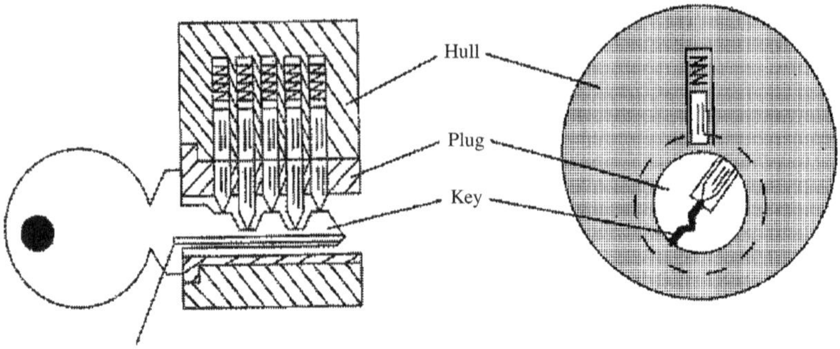

Fig. 7.1 A schematic representation of a lock-and-key.

relevance of this process to biological specificity is inevitable but misleading, as was previously argued.

Generally speaking, symmetry means sameness or indistinguishability under some transformations. A lock and a key involve symmetric transformations. When a fit is made, the key and the lock turn into a single unity that preserves its symmetry under the *rotation* of the plug. Moreover, the symmetry of the composed unit, key–lock, is possible through the symmetry of each composing unit (*i.e.* the key) and its reflected image (*i.e.* the lock). In this sense, mechanical specificity clearly involves symmetric transformations.

The symmetry, which is evident in mechanical specificity, explains the limited scope of a mechanical response. Symmetry is a matter of all or none. Either the object is symmetric (*i.e.* preserves its identity under certain transformations) or not, and when symmetry is gained a single response is produced: either the lock is opened or not. In contrast, living systems, as meaning-making systems, work according to a different logic that combines an interesting dialectic between symmetry and asymmetry. Following the work of my colleague, the philosopher Steven M. Rosen (1994), I will name this dialectic "synsymmetry".

Indeed, symmetry is evident in living systems and in different scales of analysis. Concerning bio-molecular structures, the existence of symmetry was explained by the argument that "the lowest energy state of an assembly is a symmetrical one". Indeed, "life requires rest and binding, harmony and stability" (Blundell and Srinivasan, 1996, p. 14244), but also asymmetry which is the basis of flexibility, dynamics, and change. As was quoted in Weyl's (1957) classic text,

> Symmetry signifies rest and binding, asymmetry motion and loosening, the one ... formal rigidity and constraint, the other life, play and freedom. (p. 16)

Let us take ligand-receptor binding as an example. The antibody presents a unique dialectic between order and disorder, symmetry and asymmetry. For example, textbooks present the idealized structure of the antibody as a structure with mirror image symmetry of the light and the heavy chains. However, the unique sequence of amino acids in the combining site is asymmetric. In addition, the final conformation of the binding site is determined by an interaction with a ligand that perturbs the structure of the antibody (Cohen, 2000a), and therefore, the binding site cannot be described in static terms of symmetry. Flexibility assumes the ability to move between different conformational states and, therefore, asymmetry.

In practice, the dynamics of ligand-receptor binding cannot be described in symmetric terms. The antibody may bind to an enormous number of ligands, but as suggested by Cohen (2000a), only those ligands that move the antibody to a specific conformational state may be regarded as antigens. In other words, it is not the ligand-receptor binding in itself that determines the meaning of a ligand as an antigen, but the resulting and unique conformational change that defines the ligand as an antigen! This thought-provoking suggestion invites inquiry into symmetry–asymmetry dialectics in immune specificity.

Symmetry and asymmetry of geometrical structures is not the only form of symmetry. There can be other senses of symmetry much more relevant for understanding cognitive systems like the immune system. I suggest that symmetry may have a wider interpretation with regard to cognitive tasks performed by living systems, especially with regard to meaning making. Following Piaget, I suggest that the symmetry of an object is achieved when *different perspectives converge through inferential processes to the same conclusion to yield a response with regard to the identity of the object whether a linguistic or a biological sign*. For example, object permanence is achieved when an infant infers that the object preserves its identity under various spatial transformations. That is, the symmetry of the object is *restored* through an inferential process that transcends the different perspectives from which the object is observed. This suggestion makes an inevitable link between symmetry restoration and computational reversibility as discussed in previous chapters. A similar cognitive process of symmetry restoration is evident in various forms of meaning making from comprehension of signals during animal play behavior to immune recognition. In the immune system different agents have different and limited perspectives on the signal, and in order to restore symmetry they have to "co-respond" (Cohen, 2000a, b) and communicate to achieve a global integrated view of the situation. B cells change their conformation in response to the antigen but cannot sense the context, while T cells respond to the amino acid sequence of the antigen through the major histocompatibility complex (MHC) but cannot respond to the protein's conformation. The macrophages sense the context of damaged tissue but cannot sense either the conformation or the amino acid sequence of the protein antigen. That is, each agent has a limited perspective but the symmetry-identity of the object is gained when perspectives converge and inferences are drawn. The importance of inferential processes in symmetry restoration and immune specificity brings us to the next section.

3.3. Hypothetical Inference

The antibody has a certain structure, which is perturbed by the antigen. In the appropriate context, this perturbation leads to the immune response. In this section, I would like to argue that this process should be described in terms of *abductive* or *hypothetical* inference.

The idea that the immune system is a cognitive system was suggested by Cohen (2000a), known as the author of the cognitive paradigm in immunology (Cohen, 1992). As a cognitive system the immune response involves a process of inference/reasoning whether the suspicious agent is an antigen or not. There are different types of reasoning. In deductive reasoning, conclusions are necessarily derived from premises through logical rules of inference. In inductive reasoning, conclusions are generalities that are derived from a sample of observations. These two forms of reasoning are clearly inadequate to describe the majority of inference processes in biological and cognitive systems alike. This is the reason why Peirce's idea of *abductive reasoning* is relevant to our discussion.

Peirce uses the term *habit* to describe "[readiness] to act in a certain way under given circumstances" (Pragmatism, CP 5:480, 1907). Nature is characterized by habits. The conformations of the protein molecules that comprise the binding site of an antibody follow a habit when they fold into well-known motifs. In terms of complexity sciences we may describe a habit as a *basin of attraction*. Indeed, Cohen (2000a) argues that the stable alternative shapes of a receptor protein are alternative *basins of attraction*. According to this suggestion:

> A ligand is a molecule that, through binding, can affect its receptor's conformational basin of attraction. Many sticky molecules may bind to a receptor protein, but only those that affect the response are true ligands. (Cohen, 2000a, p. 128)

According to Peirce's terminology: We may say that the binding of the antigen *perturbs* a habit. This perturbation leads to what Peirce describes as abductive inference or hypothetical inference, which is a process capable of producing "no conclusion more definite than a conjecture" (Prolegomena for an Apology to Pragmatism (MS[1] 293), NEM[2] 4:319–320, c. 1906). In other words, abductive inference is an intelligent guess or hypothesis. In the immunological context the conjecture is that the ligand is an antigen. This conjecture is examined against background signals (*i.e.* context) in a complex deliberation process between varieties of immune agents. Cohen describes this process as "co-respondence" and emphasizes its importance for better understanding the behavior of the immune system (Cohen, 2000a).

Meaning making is not a deductive process. Making sense always involve a risk, a guess, and hypotheses. Processes of inference have been studied primarily by psychologists and cognitive scientists. It may be the time to join forces with them and study processes of inference during immune recognition.

4. Conclusions

Previously, I presented a schematic illustration of lock-and-key interaction. Is there a graphical illustration that may represent biological specificity in terms of emerging meaning? One of my colleagues, Irun Cohen, challenged me with this question and after dwelling on it for a while I decided that Brancusi's famous sculpture "The Kiss" (1907) best represents biological specificity (Fig. 7.2).

Why does it represent biological specificity as a meaning-making process? Both in natural language and in biology the sign and the complimentarily (of molecules) *are not the information itself but a gate for information transfer in context*. This idea echoes Bakhtin's approach to codes. As summarized by two Bakhtin scholars:

> A code is only a technical means of transmitting information; it does not have cognitive, creating significance. (Morson and Emerson, 1990, p. 58)

In "The Kiss" we have interaction, we have specificity, which is evident from the complementarity of the two figures, but most importantly, this specificity is a gate for the flow of information (concerning passion? love?) rather than the love itself. In the same vein, structural complementarity cannot explain immune specificity. The structural complementarity is only one aspect in making sense of a signal. Converging perspectives (*i.e.* symmetry restoration)

---

[1] MS (number) refers to Peirce manuscripts.

[2] NEM (x:xxx) refers to NEM (volume:page number).

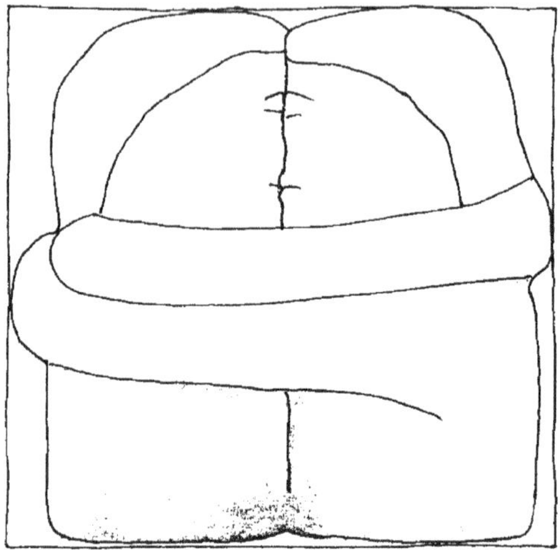

Fig. 7.2 The kiss.

through inferences are necessary for making sense in a complex environment, much more complex than the environment provided by the lock.

In sum, in this chapter, I argued that immune recognition is better described in terms of meaning making than in terms of the lock-and-key metaphor. This argument should be judged on theoretical and practical grounds alike. On theoretical grounds, it was argued that the lock-and-key metaphor is not only inappropriate for describing immune specificity but positively misleading (Carneiro and Stewart, 1994). This argument should not be taken to the extreme. The lock-and-key metaphor may help us understand ligand-receptor binding in the mature phase of the immunoglobulin although it cannot fully explain the specificity enigma.

Meaning making, rather than an appropriate metaphor for immune specificity, is an alternative way of conceptualizing immune specificity. In scientific work, conceptualization should be preferred to metaphors, especially if this conceptualization addresses the difficulties introduced by an existing metaphor. This argument seems to apply to the lock-and-key metaphor and the alternative conceptualization of immune specificity as a meaning-making process.

To review, Efroni and Cohen (2003) argue that a good biological theory is one that serves the process of discovery and opens the way to "otherwise unthinkable research". The inevitable question is whether meaning making can serve the process of discovery by opening new paths of inquiry. Let us discuss a few new research questions that emerge from the conceptualization presented in this chapter.

If the immune system is a cognitive system, as suggested by Cohen, and if immune specificity involves hypothetical inferences, as was argued above, how can we study or simulate this process of inference? Although Efroni, Harel, and Cohen (2003) have recently introduced a new methodology for the dynamic modeling of the immune system, the inference system underlying immune specificity deserves special treatment with emphasis on the identification of the relevant background signals (*i.e.* context) on which the hypothesis is examined. Second, it was argued that a recursive-hierarchy underlies the existence of immune specificity. However, it is not quite clear how cross-scale interactions work. Simulations of complex systems frequently deal with bottom-up processes and it is not quite clear how different layers of a biological system interact to achieve a specific response. Although Conrad was making the first moves to address this question our knowledge of immunology as a recursive-hierarchical system is still in an embryonic phase. Without any theoretical progress in understanding those systems no real advance can be made on the specificity enigma. This is definitely a challenge facing future research of biological systems in general and the immune system in particular.

In this chapter, I also showed that a semiotic approach to immune specificity might be an alternative to the dominant mechanistic-reductionist perspective. In the next chapter I follow this path and delve more deeply into a semiotic approach to immunology and illustrate the way this approach may shed new light on the issue of self and non-self discrimination.

Chapter 8

A Point for Thought: Reflections on the Immune Self

Summary

The immune self is one of the main organizing concepts in immunology. However, it is not quite clear what the immune self is and heated debates have taken place among immunologists concerning the meaning and usefulness of this concept. In this chapter, I further illustrate the benefits of a meaning-making perspective and argue that the problem of the immune self can be approached as analogous to the problem of finding the different senses of the sign in semiotics. Following this suggestion, I would like to present the idea that the immune system is a meaning-making system and in this context to provide a novel conceptualization of the immune self that integrates several ideas from immunology and semiotics.

1. Introduction

Immunology has been described as the science of self and non-self discrimination (Abbas *et al.*, 2000) but the meaning of the *immune self* has been a disputed issue. What is the immune self and why has the concept of self been introduced to immunology at all?

The concept of self is traditionally associated with disciplines such as philosophy and psychology. Indeed, there is a whole branch of psychology known as *self psychology* and in philosophy the concept of self was lengthily discussed with regard to the question of personal identity. However, my aim is not to discuss the self as a property of human beings and as a response to the questions: "Who am I?" or what is the stable essence of my identity. My aim is to discuss the "self" of the immune system. Are those two different selves? Is the self a concept that is applicable both to philosophy/psychology and immunology? Is the self a metaphor that was imposed on the immune system or is it a crucial organizing concept for theoretical immunology?

As suggested by Howes (1998), there are number of "fascinating parallels that might be drawn between theoretical developments concerning self in philosophy and in immunology" (p. 1). These parallels cannot be denied. The concept of self was introduced to immunology by Sir Frank Macfarlane

Burnet after it has been intensively elaborated in philosophy and psychology. As an imported metaphor from other fields of inquiry the concept was inevitably loaded with associations and connotations that have clearly influenced the idea of the immune self. Although interesting parallels exist between the concept of self in the humanities and the concept of self in immunology, these parallels should not blur the significant differences between the self-concept as it is used in these distinct fields of inquiry. This point was raised by Tauber (1998):

> The immune system is not a human category. We make a category error in assigning human descriptions to lymphocytes and antibodies, which reside in their own domain, objectified, possessing no self-consciousness as we understand our own psyches, and hopefully freed from our possessive prejudices. (p. 8)

Tauber's comment should make us sensitive to our use of the concept of self in immunology but on the other hand it should not discourage us from discussing its meaning in a critical and reflective way. My advice is that, as a first step in our inquiry into the immune self, we should avoid the variety of senses, connotations, and associations of the concept of self as it was discussed in philosophy and related disciplines, and face the issue of the immune self from a different perspective. Instead of surveying various philosophical definitions of the self, I prefer to start with those who *condemn* it. Who are those who approach the concept of self (whatever it is) with negative feelings? The answer is the *mystics*.

In the *Encyclopedia of Mysticism* (Ferguson, 1976) the self is described as a "barrier to the highest" (p. 167). From the *Theologica Germanica* edited by Luther in the 16th century to the writings of the mystic Meister Eckhart, the self is blamed for being a barrier to the mergence of man with the totality named *God.* The ultimate solution to this barrier is known in nature as death. This is the reason why so many mystics do not see death as such a horrible incident. After all, merging with the totality of the universe in one way or another is the mystic's ultimate aim. There are other perspectives of course. Woody Allen said once: "It's not that I'm afraid to die. I just don't want to be there when it happens". In nature, organisms are closer in mind to Woody Allen than to Meister Eckhart. This is not a mere philosophical preference. Organisms are not so fond of death and they will do their best to constitute their differentiated existence and to delay their encounter with the totality of the universe. In other words, the self in biology is no other than:

> *The organism's systemic closure defines it for all practical reasons as a differentiated unit of activity/analysis*

This working-definition should not be confused with an explanation of what the self is, because it does not explain to us what the processes that constitute this systemic closure are. However, this working-definition provides us with a good starting point that avoids the irrelevant senses of the concept of self as it was discussed in philosophy and related disciplines. This working-definition also explains to us why despite all the accompanying difficulties the "self" is still a major organizing concept in immunology, which is heatedly debated by the theoreticians in the field. Organisms struggle to constitute their systemic closure (*i.e.* their self) and the immune system is one of the crucial systems for fulfilling this function.

In this chapter, I do not aim to provide a complete survey of the literature dealing with the immune self or to review its different senses or historical and philosophical origins, a task already done by Tauber (1996). My aim in this chapter is much more restricted in focus. I would like to present the argument that the problem of the immune self is analogous to the problem of finding the meanings of the sign in semiotics. Following this line of reasoning, I would like to elaborate on the idea previously introduced, that the immune system is a meaning-making system, and in this context to provide a novel conceptualization of the immune self that integrates several ideas from immunology and semiotics.

In this context, it is worth asking questions such as what is the meaning of the immune self and whether it is a meaningful concept that is really crucial for immunology. In other words, a basic question is whether we really need the concept in order to understand the immune system. Surprisingly, this fundamental question is still debated in immunology. There are some researchers who answer this question with a categorical *yes* or *no*. For example, according to Langman and Cohn (2000) the answer to the above question is "Yes!" To quote:

> Any biodestructive defense mechanism must distinguish between the 'bio' of self (*e.g.* host) and the 'bio' of non-self (*e.g.* pathogen). (p. 189)

This categorical answer is not satisfactory since it answers the question with its own terms explaining the "Yes" by the same terms (self and non-self) it supposes to explain. However, let us start from a highly popular model in immunology that answers the above question with a categorical *no*. This model—the danger model—dismisses the concept of self in immunology and offers an alternative. I would like to critically examine this model before delving into the different notions of the immune self.

2. Danger!

Polly Matzinger (2002) has offered the *danger model* to cope with some of the most bothersome questions concerning self and non-self discrimination in immunology. In fact she argues that the self and non-self discrimination itself

> can be removed and replaced with the view that the body cares more about what is dangerous (damaging, toxic, etc.) than what is non-self. (Anderson and Matzinger, 2000, p. 232)

The question that immediately pops into mind is: dangerous to whom? Danger is not an abstract concept. Danger is always a danger for someone or for something and if this something is the self then the argument is problematic. That is, barring the immune self from the main entrance for the danger model is just a way of inviting the immune self in through the back door. Nevertheless, Anderson and Matzinger (2000) made an attempt to avoid the self and non-self terminology. Their line of reasoning is as follows. When the immune system encounters an antigen it should first decide whether to respond. If the answer to this question is positive then there are three more questions: (1) how strongly to respond; (2) with what effector class; and (3) where? In this decision-making context, the major question is: "How can I do this without destroying the tissues that I am meant to protect?" (p. 231). Please note that again there is a clear philosophical fallacy in this presentation that aims to get rid of the concept of self. The questions arise when the immune system is faced with an antigen. But what is an antigen? The meaning of an antigen is embedded in the general idea of self and non-self discrimination, so the argument cannot avoid circularity.

The danger model suggests that antigen-presenting cells (APC) are being activated by danger signals from injured cells. The APCs co-stimulate T helper cells that support B cells attached to the antigen. It is further argued that "any intracellular product could potentially be a danger signal when released" (p. 302). Following this suggestion the model pretends to explain the mother's tolerance for her fetus and the intolerance of the immune system to grafts. It has been argued that the fetus is not rejected by the body since it does not send alarm signals, and, that transplants are rejected due to the alarm signals released during the surgical intervention (p. 304).

For some people the danger model is appealing in its simplicity and commonsensical approach. However, it is accompanied with severe theoretical difficulties. Vance (2000) has poignantly presented some of these difficulties. Let me present some of them. First, the idea of danger is not new to immunology (Janeway, 1992). Immunologists usually adopt a two-signal model of lymphocytes activation (Vance, 2000). The first signal (signal one) is received through the T-cell receptor (TCR). This signal functions to discriminate between self and non-self. Signal two or *costimulation* is provided during genuine infection through APC. So what is the new gospel in bringing signals of infection (*i.e.* danger) to the forefront of immunological theory? It seems that the answer lies in the ambition of the danger theory to allegedly create a macro-level theory of immunology that avoids the language of self and non-self discrimination. In other words, the danger theory seeks to exclude the concept of the immune self and to replace it with the concept of danger as a new constituting concept for immunological theory, but with the accompanying difficulties previously presented. Let us present more difficulties.

The first is that the danger theory attacks the immune self as *straw man* without considering it as what it really is: a helpful heuristic. According to Vance the immune self is "a useful heuristic device" and not the "whole story" of the immune system. In this context there is no point in attacking the straw man of the immune self and replacing it with the concept of danger. Indeed, the concept of the immune self is (1) ill defined, (2) has many exceptions, and (3) does not account for a wide enough range of immunological phenomena. However, these are also difficulties that characterize the concept of danger. As correctly argued by Vance, a theory is a generalization and as such anomalies and exceptions abound. In fact this property is what characterizes a scientific theory. Only pseudo science can explain everything.

As Vance further argues, the meaning of danger in the danger theory is problematic. According to danger theory signals of danger are not transmitted by infectious agents themselves but by the host *tissues* that are damaged in the course of infection. Thus exogenous signs of infection (such as bacterial DNA) are excluded from the category of danger. Moreover, the conflation of danger signals and inflammation is problematic. Inflammation is mediated by immune effectors. In fact, inflammation is defined as a localized protective response that results from injury or destruction of tissues. Therefore inflammation is both the *result* of immune response and according to the danger theory a *cause* of immune response through the signals of danger. This circular argument is a problem. In sum, the most pretentious attempt to dismiss the immune self as a major organizing concept in immunology seems to suffer from severe theoretical difficulties. Therefore, we cannot avoid the task of clarifying the meaning of the immune self. The next section presents the genetic-reductionist interpretation of the immune self.

3. A Reductionist Explanation of the Immune Self

How do we know how to differentiate between self and non-self? The genetic-reductionist approach suggests that there is a single genetic criterion for self-identification. It is a genetic fingerprint that allows the immune system to differentiate between the self and the non-self. Remember the

MHC? It first appeared as if this genetic marker might provide us with the ultimate criterion for self and non-self differentiation. Rolston clearly expresses this idea:

> The recognition of the non-self is signaled by the molecules of the major histocompatability complex (MHC). Class I molecules are placed on every nucleated cell in the body to identify the self. It is also important to discriminate which cells to kill, and this is done by T cells, using class II molecules, which are placed, in macrophages, B cells and some T cells. (Rolston, 1996, quoted in Howes, 1998, p. 3)

It should be noted that according to this suggestion the non-self is an empty slot. There is only a self, which is identified through the genetic marker of the MHC. An entity is recognized as non-self only if it lacks the sign of self. In other words, the foreignness of the antigen is implied by not having a self-marker.

The idea of an *identity marker* or an identity sign has interesting cultural roots. What are the identity markers we are familiar with? The first answer popping into my mind is the fingerprint. As someone who as a child had great pleasure in reading detective stories, I learned to appreciate the importance of this identity marker in solving crimes. The importance of the fingerprint in police work was specifically evident in a period in which criminals were not familiar with this incriminating sign. Later, criminals learned how to use gloves and the crime scene has since turned into a much more difficult text to interpret.

The fingerprint has been always portrayed in my imagination as a solid identity marker and as undisputed forensic evidence. Not only in my imagination, but also in forensic work, this identity marker has been considered an undisputed source of information. The question of course is: How do you know? How do you know that the fingerprint is an undisputed identity marker? My own naïve reaction to this question would have probably been: I assume that the fingerprint was scientifically tested and approved as an undisputed identity marker. Surprisingly:

> The underlying scientific basis of fingerprint individuality has not been ruinously studied or tested. In March 2000, the U.S. Department of Justice admitted that no such testing has been done and acknowledged the need for such a study. (Pankanti *et al.*, 2002, p. 3)

Several years passed since the U.S. Department of Justice released their statement and progress has been made in establishing the scientific basis of the fingerprint. This incidence is not just a philosophical or historical anecdote. For years the scientific basis of this identity marker was not well established. Nevertheless the fingerprint has been conceived beyond any criticism as scientifically based and as the ultimate identity marker. This case should raise our awareness of the different meanings of identity markers to include the genetic markers of the self.

The genetic-reductionist approach clearly corresponds to the *referential* theories of meaning in semiotics that aim to explain the meaning of a sign through a simple correspondence with a reference. The MHC seems to be a sign that clearly corresponds to the "self" whatever it is.

Following Frege, we should differentiate between *sense*—the semantic content of the sign—and its reference—what the sign refers to. The MHC does not seem to have a simple and concrete reference. It does not point to a concrete object. However, it has a sense. It is a sign of the self. What is the self? Here we enter a dead-end alley. The genetic-reductionist approach does not explain to us what the immune self is: Our complete genome?; The codable part of our genome? It just points at MHC as a sign corresponding to the self the same as the sign *cat* corresponds to the member of the feline family that is currently sitting near me while I write this text.

The genetic-reductionist approach is not totally wrong just as referential theories of meaning are not totally wrong. Meaning can be interpreted to a certain extent and in certain cases in terms of a correspondence between a sign and a signified. When a child learns to use her language by pointing at a cat and saying "cat", a direct correspondence is established between the sign and the signified. The problem is that in complex systems the meaning of a sign such as "immune self" *cannot be exhausted* by using the simple means of direct correspondence between a sign and a signified. As summarized with regards to immunology (Tauber, 1998, p. 458):

> Although an understanding of such immune behavior canonically begins with the major histocompatability complex (MHC), *its complete characterization* appears to reside at levels of biological organization beyond the gene. (Tauber, 1998, p. 458; emphasis mine)

Let me support Tauber's argument by pointing at the theoretical and empirical difficulties with the genetic-reductionist approach to the immune self. The first theoretical problem is that in nature a *self* is a dynamic object. One does not have to be an expert in the dynamics of biological systems in order to recognize this fact. If the self is recognized through a strict and single genetic criterion how is it possible to explain the changes in the identity of organisms through evolution? This question remains unanswered if we adopt a simplistic reductionist approach to the immune self. The same problem is evident in semiotics. Indeed, a sign may correspond to the member of the feline family that is currently sitting on my lap. However, the sign *cat* may also signify a Jazz player or a skillful burglar who climbs like a cat. A simple, static and permanent correspondence does not exist between a sign and a signified.

When examining empirical evidence, things become even more problematic. Tolerance and autoimmunity are two phenomena central to understanding the problems of the genetic-reductionist approach. Autoimmunity is a process in which the immune system turns against constituents of the host that it is supposed to defend. As I will show later this definition is oversimplified but for instructional reasons let us accept it for the moment. Autoimmunity is usually associated with disease. For example, lupus is a kind of autoimmune disease in which the antibodies identify the host tissues as "non-self" and might cause arthritis and kidney damage. However, autoimmunity is not necessarily a pathologic process. For example, it has been argued that

> autoimmune T cells that are specific for a component of myelin can protect CNS neurons from the catastrophic secondary degeneration, which extends traumatic lesions to adjacent CNS areas that did not suffer direct damage. (Schwartz *et al.*, 1999, p. 295)

The implications of this argument for the genetic-reductionist conception of the self are clear. The self is not a stable and well-defined entity, which is protected from the non-self through the immune system but a contextual and dynamic construct. The immune system may turn against host constituents as a normative function of bodily maintenance. As argued by Cohen (2000b, p. 215) with reference to the context of inflammation:

> The difference between autoimmune protection and autoimmune disease, it appears, is a matter of intensity and the timing of the autoimmune inflammation.

That is, autoimmunity is not always a simple matter of an immune system that attacks the self it is supposed to protect. It is not a self versus itself.

Tolerance is the complementary aspect of autoimmunity. It concerns the immune system's ability to ignore its own constituents *even if these constituents do not bear the genetic identity marker of the "self"*. My example concerns a bacterium by the name of *Escherichia coli*. When this bacterium is found in high concentrations in food it is an indication to the low-hygienic standards of the restaurant, and the restaurant owner might lose his license. However, this bacterium rests peacefully in our colon, as well as in the normal flora of the mouth without being aggressively attacked by our immune system. This bacterium is clearly not a part of the self as defined by the genetic-reductionist approach.

Tolerance of parasites is not an exception in nature and *E. coli* is just a specific instance. Sometimes, like talented imposters, intruders into the host self develop unique machinery to hide their identities. However, in many other cases they are simply tolerated by the host. Organisms, human beings for example, host a variety of parasites that live in perfect symbiosis with them. These parasites, such as *E. coli* are not a part of the self in the genetic sense. However, during the evolution of mammals parasitic relationships have been established with this bacterium to produce vitamins B12 and K, and to aid the digestion process. In sum, the immune system tolerates the presence of the *E. coli*, a fact that the genetic-reductionist approach to the immune self may find difficult to explain.

Let me give you another simple example of immune tolerance to cells that clearly do not have the genetic marker of the self. A woman having sex with a man hosts in her womb his sperm cells. Assuming that her partner is not her twin brother, his sperm cells clearly do not carry the marker of her self. How is it that the host immune system does not destroy the sperm as a non-self? And when the fertilized egg develops into a fetus how does the immune system tolerate the fetus?

As suggested by Medawar (quoted in Choudhury and Knapp, 2000) the fetus represents an immunologically foreign graft that is maternally tolerated during pregnancy. Medawar suggested three hypotheses to explain this phenomenon (Mellor and Munn, 2000): (1) physical separation of mother and fetus; (2) antigenic immaturity of the fetus; and (3) immunological inertness of the mother. However, it is clear that no single mechanism resolves the quandary.

How does the mother's immune system tolerate the fertilized egg? This is an interesting, and to a large extent, unanswered question in the biology of reproduction. However, there is an interesting finding that should be mentioned in this context. In an article entitled: "*Sex is Good for You*" (Buckland, 2002) interesting and relevant research was reviewed. In this research, it was argued that recreational sex—sex with no procreational purpose—can have a positive impact on pregnancy. But an important qualification should be added: *sex with the same partner*. Sex, early, often, and with the intended father may help overcome the reluctance of the mother's immune system to accept a fetus that expresses foreign proteins from the father's genes. That is, the more accustomed the women's immune system to the father's sperm, the more habitual the encounter, the less likely her body will be to reject the fetus.

From this research we learn two things. First, the Catholic Church is found once again to misunderstand the nature of living organisms. If pregnancy is encouraged then recreational sex with the same partner should be encouraged too. The second lesson is that the *somatic* aspect of the immune system is crucial for understanding a variety of immunological phenomena. Not everything can be reduced to the genes. There are things we should learn by ourselves and learning is a built-in characteristic of any intelligent system like the immune system.

Along the same lines, sperm proteins arise after the development of neonatal immune tolerance, that is, after the immune tolerance has basically been established. It is known that crude sperm proteins are highly immunogenic in all species (McLachlan, 2002). How is it that these non-self cells are tolerated by the host? What is the mechanism that prevents the generation of sperm antibodies (SpAb)? This is not a theoretical question since a significant portion of infertility[1] among men is attributed to the generation of SpAb. Indeed, it was found that approximately 6% of male infertility problems are the result of autoimmunity, that is, infertility that is caused by the immune system that identifies the sperm cells as antigens.

Antisperm antibodies (ASA) can be produced by both sexes (Shulman, 1995). Males can create autoantibodies against their own spermatoza and females can produce antibodies against a male's spermatozoa. ASA are found in semen (*i.e.* the viscous whitish secretion of the male reproduction organs), sera (*i.e.* plural of serum, watery fluid that contains antigens), or bound to outer sperm membrane. In woman ASA are found in blood, ovarian, follicular fluid, and vaginal or cervical secretions (Choudhury and Knapp, 2000). The precise mechanism in which ASA affects fertility remains questionable (Choudhury and Knapp, 2000).

Again we should ask ourselves how the body tolerates the emerging sperm cells. Why doesn't it usually attack them as non-self? This is a basic question and as we try to answer it our ignorance is exposed. The answer to the above question is disappointing:

> Attempts to identify a universal or clinically relevant antigen–antibody interaction associated with infertility has thus far been unsuccessful. (McLachlan, 2002, p. 37)

A lack of knowledge is usually not a proof for an argument. However, in our case the lack of knowledge concerning sperm tolerance or autoimmunity is

[1] Due to the several senses of infertility, I use infertility in the sense of a failure to conceive after frequent unprotected intercourse.

an indication that the genetic-reductionist approach to the immune self is oversimplistic and cannot provide us with the answers we are looking for. It is time to move on to another perspective of the immune self.

4. Burnet and Saussure

The idea of self in immunology appears first in Burnet's 1940 *Biological Aspects of Infectious Disease* (Crist and Tauber, 1999). The concept was first introduced in the context of an amoeba digesting its prey:

> If we are to describe and discuss such phenomena scientifically, we must for the present at least be satisfied with a ... biological approach. Is there any simpler way of looking at this relationship between the eater and the eaten? It may be that something useful can be gained by concentrating on the most obvious aspect of all—that the engulfed micro-organism is not the amoeba itself, The fact that one is digested, the other not, demands that in some way or other the living substance of the amoeba can be distinguished between the chemical structure characteristics of the "self" and any sufficiently different chemical structure which is recognized as "non-self." Here we seem to have an important general character on animal protoplasm which may provide a connecting thread to help link up some of the very diverse manifestations of the defense processes which we shall have to consider. For with one very important exception, every disease-producing invasion of the body is by some type of organism whose intimate structure is foreign to the body. All such invasion can, in at least a proportion of instances, be overcome by natural processes. Perhaps it is significant that when invasion by the uncoordinated growth of the body's own cells (cancer) occurs, natural processes never succeed in overcoming it. (Burnet, 1940, p. 29, quoted in Crist and Tauber 1999, p. 520)

By introducing the metaphor of self, Burnet has clearly offered a new heuristic rather than uncovering a biological *Tera Incognita*. The fact that he chose to put the concept of "self" in quotation marks indicates the heuristic nature of this move. As I previously explained the biological self is no more than our way of conceptualizing the systemic closure of the organism. Burnet has clearly used this basic sense of the concept when he introduced it in the specific context of amoeba digesting its prey. However what is the relation between the general notion of a biological self and the immune self? Burnet's perspective on the immune self can be comprehended from his

Clonal Selection Theory (CST). However, the relation between Burnet's CST and self and non-self discrimination is far from being simple.

It was argued by Silverstein (2002) that the difference between the central hypothesis of CST and the subsidiary hypotheses, such as the one concerning self and non-self discrimination, has been confused. According to Silverstein (2002) the core of the CST involves four hypotheses:

1. The entire immunological repertoire develops spontaneously in the host.
2. Each antibody pattern is the specific product of a cell and that product is presented on the cell surface.
3. An antigen reacts with any cells carrying its specific receptor to signal-cell proliferation and differentiation.
4. Some of these daughter cells differentiate to form clones of antibodies, whereas others survive as clones of undifferentiated memory cells.

These hypotheses say nothing about the nature of the immune self and the way self and non-self discrimination is conducted. However, this argument is theoretically weak due to the importance of the immune self in Burnet's theory and the meaning of the immune self as implied from the CST. Let me explain Burnet's conception of the immune self while using CST as the context of interpretation.

The genetic-reductionist approach suggests that there is only a self, signified by a genetic marker. From this theoretical position it is implied that the non-self is not an actual entity but a synonym for a genetic foreigner. Burnet presents a mirror-image perspective in his CST. It was suggested by Burnet that lymphocytes with reactivity against host constituents are destroyed during development, and only those lymphocytes that are non-reactive would be left to engage the antigens of the foreign universe. The foreign is destroyed by immune cells and their products, whereas the normal constituents of the organism are ignored. That is the immune system recognizes only the non-self and the self is an empty term. Burnet's CST explains from a very simple evolutionary perspective why we tolerate ourselves. We tolerate ourselves because those that were unable to differentiate between self and non-self simply did not survive.

There are severe difficulties with Burnet's conception of the immune self, for example, the fact that self-recognition is clearly evident in the immune system. I will point to these difficulties in the next sections, but in the current phase of our analysis, I would like to point to some similarities between Burnet's conception of the self and Saussure's conception of the sign.

Burnet's concept of the self is purely differential and negative. The self exists only as a background for the identification of the foreign, of the non-self. In a certain sense, this position is similar to the one presented by

Ferdinand de Saussure in his classical text *Course in General Linguistics*. According to Saussure:

> *In the language itself, there are only differences.* Even more important than that is the fact that, although in general a difference presupposes positive terms between which the difference holds, in language there are only differences and no *positive terms*. (Saussure, 1972, p. 118, italics in original)

I will repeat parts of this quotation several times because differences are the basic units of our analysis. What does Saussure mean when he says that in language there are "only differences"? Let me explain this statement. For Saussure language, as an abstract system of signs (*i.e. la langue*), is "a system of distinct signs corresponding to distinct ideas" (Saussure, 1972, p. 26). That is, in itself a sign means nothing. It exists solely by being differentiated. According to this interpretation the sign *cat* has no intrinsic meaning. The "catness" of the cat is not embedded either in the way *cat* is pronounced or in the concept of cat. The same is true for our self. During our lifetime our self significantly changes: cells die and are replaced by new ones, our mental content changes during our development, and so on. There is nothing that is intrinsic in our self that may define us as the same person through the years. According to this line of reasoning, our identity is primarily and negatively established by our differentiation from others. To use an analogy from mathematics (Kempe, 1886), a pair of points consists of units that, taken in isolation, are indistinguishable. Each unit in this pair is distinguished only by holding a differentiated position from the other.

Saussure's statement is applied to the sign as an isolated unit which is "purely differential and negative" (Saussure, 1972, p. 118) as a phonetic or a conceptual unit. To review, for Saussure the meaning of a word is the "counterpart of a sound pattern" (p. 112). In this sense the meaning of the sign *cat* is its corresponding concept of cat. Saussure suggests that *meaning* should be distinguished from *value*, which is important for understanding the abstract nature of any system of signs. This idea will be repeated in the book and therefore the reader should keep it in mind.

A value involves: "(1) something *dissimilar* which can be exchanged for the item whose value is under consideration and (2) *similar* things which can be *compared* with the item whose value is under consideration" (p. 113). For example, money is an abstract system of signs/values. In this system, like in the linguistic system, a one-dollar bill has no meaning in itself. The meaning of a one-dollar bill can be determined only in a closed system of values. To determine the value of $1 we should know that a one-dollar bill can be exchanged for something different (*e.g.* a candy bar) and that its value can be compared to another value within the same system of currency (*e.g.* exchanging it for Euros). The language system is a system of pure values whose function is to combine the two orders of difference—phonic and conceptual—in the making of signs.

Turning to immunology, the similarities are clear: the immune self has no meaning in itself. The immune self is only negatively established through the existence of the other—non-self. However, where Burnet stops his analysis, Saussure presents a systems-oriented approach, moving from the isolated sign of language as an abstract system, and pointing to the social semiotic dynamics that makes this abstract system of values materialize in practice.

Surprisingly, Saussure theory of language as a *social network of signs* brings us, once again, to the idea of self and non-self discrimination. As suggested by a semiotician without any reference to immunology:

> Meaning is an embodied relation between self and non-self on the basis of the individual's entraining into the higher-order and transindividual structures and relations of langue. (Thibault, 2005)

Translating this poetic paragraph into simple words this excerpt means that it is only by going beyond the individual level of analysis and entering the semiotic matrix that the relation between self and non-self can be clarified. As will be explained in the next section, this approach may take the place of the classical conception of the self as a single monolithic entity. This conclusion naturally brings us to Jerne and his network theory of the immune system.

5. Jerne and Peirce

> I think there is now a need for a novel and fundamental idea that may give a new look to immunological theory. (Jerne, 1974, p. 380)

An interesting alternative to Burnet's concept of the self was suggested by the Nobel Laureate N. K. Jerne in a discussion of his network theory of the immune system (Jerne, 1974). This theory clearly corresponds to Saussure's idea and pushes it to its limits within immunological theory.

Jerne suggests that the "progress of ideas" in immunology follows a path from application (*i.e.* vaccination), through description (for example of antibodies), mechanisms (*e.g.* selection clones), up to systems analysis of network cooperation, and suppression of immune agents. He posits his theory in the final phase of this progress and approaches the immune system using the network metaphor. Before presenting the gist of his theory let us clarify some of the terms that he uses.

An *antigenic determinant* is a term that denotes a single antigenic site or epitope on a complex antigenic molecule or particle. Jerne replaces the term antigenic determinant with the term *epitope.* He also replaces the term *antibody combining site* with the term *paratope.* In this sense the paratope is complementary to the epitope. Next he introduces the terms *allotype* and *idiotope.* Allotypes are "Antigenic determinants that are present in allelic (alternate genetic) forms. When used in association with immunoglobulin, allotypes describe allelic variants of immunoglobulins detected by antibodies raised between members of the same species" (Jerne, 1974, p. 380). Idiotopes are "the combined antigenic determinants found on antibodies of an individual that are directed at a particular antigen; such antigenic determinants are found only in the variable region" (p. 380). In other words, an idiotope is a set of epitopes. The term *repertoire* is used to consider the repertoire of antibody combining sites or the total number of different paratopes in the immune system. By using this terminology Jerne (1974) suggests that the

> immune system is an enormous and complex network of paratopes that recognize sets of idiotopes, and of idiotopes that are recognized by sets of paratopes. (p. 381)

According to this suggestion antibody molecules can recognize as well as be recognized. This situation raises a question: What happens to a lymphocyte when its idiotopes are recognized by the paratopes (*e.g.* of another cell)? Jerne's suggestion is that the lymphocyte is then repressed. Stressing the importance of repression he suggests that the "essence of the immune system is the repression of its lymphocytes" (p. 382). This is a radical statement since it suggests that the immune system is a closed system which is primarily oriented toward itself rather than toward the destruction of foreign invaders. In other words, the system is "complete onto itself" (Bersini, 2003). The idea of a system complete onto itself is a natural derivation of avoiding a direct encounter with the relation between a sign and a signified. Unable to explain the relation between a sign (*i.e.* an antigen) and a signified (*i.e.* non-self) a dangerous tendency is to deny the existence of a signified (*i.e.* the immune self and non-self) while assuring the autonomous realm of a sign system. The problem in this case is to explain the role of the real world which is external to the system of signs.

Tauber (1997) describes Jerne as the "true author of the cognitive immune model" (p. 424) meaning that the immune system is designed to know itself. In this context, the antigens are interpreted as stimuli that cause *perturbations* in the network. There is no non-self and therefore not even a self but just a "source" of perturbation that causes the network to reorganize itself in order to restore a lost equilibrium. As summarized in Tauber (2002):

> In the Jernian network, "foreign" is defined as perturbation of the system above a certain threshold. Only as observers do we designate "self" and "non-self". From the immune system's perspective it only knows itself.

And in another place he further explains this perspective:

> Antigenicity is only a question of degree, where "self" evokes one kind of response, and the "foreign" another, based not on its intrinsic foreignness but, rather, because the immune system sees that foreign antigen in the context of invasion or degeneracy. (Tauber, 1997, p. 425)

The reader familiar with the work of Humberto Maturana and Francisco Varela (1992) may immediately recognize the similarity between their theory and Jerne's perspective. Both were inspired by the system metaphor and both promote the notion of an autonomous system, which is "closed" and subject to perturbations only. Indeed, as argued by Vaz and Varela (1978):

> All immune events are understood as a form of self-recognition, and whatever falls outside this domain, shaped by genetics and ontogeny, is simply nonsensical. (p. 231)

The problem with the network theory originated by Jerne and advanced by his proponents is that it suffers from conceptual obscurity regarding the way in which meaning is established in a closed system. The key term for understanding this difficulty is the *hall of mirrors*. Let us first read Jerne and then explain this difficulty:

> The immune system (like the brain) reflects first ourselves, then produces a reflection of this reflection, and that subsequently it reflects the outside world: *a hall of mirrors*. The second mirror images (*i.e.* stable anti-idiotypic elements) may well be more complex than the first images (*i.e.* anti-self). Both give rise to distortions (*e.g.* mutations, gene rearrangements) permitting the recognition of non-self. The mirror images of the outside world, however, do not have permanency in the genome. Every individual must start with self. (Jerne, 1984, p. 5; emphasis mine)

Jerne's use of the term *hall of mirrors* is not an intellectual whim and corresponds to an established position concerning the relation between a sign and a signified. Rosen explains this position as follows:

> Structuralist semioticians like Saussure still sought to preserve the invariance of the link between the given signifier and what it signifies. The problem is that, once classical signification is surpassed by signifying the signifier, the door is opened to an infinite regress. For now, it seems that no signifier is exempted from mutation into that which is signified. A new signifier is presumably needed to signify what *had* been the signifier, but this new signifier is subject to signification by a still newer signifier, and so on *ad infinitum*. And each time the tacit operation of the signifier is undermined by being explicitly signified, the functioning of what had been signified by that signifier is also affected. Ultimately then, we have in this "hall of mirrors" neither signifier nor signified in any stable, abidingly meaningful form. (Rosen, 2004, p. 38)

Rosen attributes this position to Derrida, but one may also find it in a sophisticated and constructive form in the semiotic theory of C. S. Peirce. For Peirce a sign is "a Medium for the communication of a Form" (MS 793 [On Signs], n.d., p. 1). In this sense, it is a member of a triad and holds a mediating position between an *object* (*i.e.* anything that we can think, *i.e.* anything we can talk about" (MS 966 [Reflections on Real and Unreal Objects], n.d. http://www.helsinki.fi/science/commens/dictionary.html) and an *interpretant* the effect of a sign on someone who reads or comprehends it. This triad of the object, sign, and interpretant is the indivisible unit of *semiosis*—an action or influence that cannot be reduced to direct encounter between pairs such as an agent and an object. In other words, any sign-mediated activity is semiosis and is triadic due to, in Peirce's words: "This tri-relative influence not being in any way resolvable into actions between pairs" (EP 2:411[2]). The process of semiosis is irreducible but ever expanding since the interpretant exists as long it is a part of a dynamic process of semiosis. Let me explain this point.

According to Peirce *meaning* is that which a sign conveys:

> In fact, it is nothing but the representation itself conceived as stripped of irrelevant clothing. But this clothing can never be

[2] EP (x:xxx) refers to EP (volume:page number).

> completely stripped off; it is only enacted for something more diaphanous. So there is an infinite regression here. (CP 1:339)

The meaning of the interpretant-self is therefore "nothing but another representation" (CP 1:339). In other words, "like the signs in general, the self manifests a trinary character. Every self, in collaboration with its signs, addresses itself to some other" (CP 5:252). The self is mediated and inferred, and like all signs must be related to otherness. The relevance to self and non-self discrimination in immunology is implied as has been described in a semiotic context:

> That is, the self, upon inferring itself into existence, sets itself apart from everything else in order that there may be a distinction between something and something else. (Merrell, 1997, p. 57)

In sum, the self comes into being by a process of semiosis. It is not a construct that is given *a priori.*

We can see that Jerne network theory clearly corresponds to Peirce's theory of semiosis. However, Peirce theory may shed light on Jerne's ideas and add depth to his network conceptualization. For example, in Peirce's sense a *perturbation* of the system is a break in a *habit* where habit is used in the sense of *regularity*. This perturbation in the process of semiosis results in an effort to reorganize the system and to restore the lost equilibrium. According to this interpretation the immune network is not absolutely autonomous. It is context-sensitive and attunes itself to perturbations; violations of habits/regulation, which we may *post hoc* define as non-self. In other words, the self may be considered as

> *The regularity of relations and interactions that constitute the systemic closure of the organism*

A disturbance to this regularity (local or global), whether it emerges from inside or outside the system, may be responded to by the immune system and defined as non-self. This interpretation preserves the flexibly dynamic and commonsensical notion of the self and at the same time explains the case sensitivity and the contextual nature of immune activity. This interpretation of the Jernian network brings it closer to the contextualist approach propagated by Irun Cohen. The next section presents the contextualist approach.

6. Cohen and Volosinov

Another response to Burnet's dominance comes from the contextual theory of immunology suggested by Irun Cohen. Cohen (1994) discusses Burnet's conception of self and non-self discrimination while using the analogy of a figure/subject and a background. According to this analogy, Burnet's theory considers the foreign as the figure/subject and the self as the background. As he explains:

> According to clonal selection, only the picture of non-self has substance; the picture of the self must be virtual. The immunological self can exist legitimately only as that which bounds the foreign. (Cohen, 1994, p. 11)

Cohen argues that this conception of the immunological self is wrong because the immune system knows to recognize the self:

> Healthy immune systems are replete with T and B cells that recognize self-antigens. (p. 11)

While genetic reductionists suggest that only the self really exists, and while Burnet suggests that the only thing that really exists is the non-self, Cohen suggests that the self and the non-self are *complementary*. He discusses this idea by using four titles: (1) substance, (2) essence, (3) origins, and (4) harmony.

The title of "substance" concerns the fact that "self-antigens and foreign antigens are made of similar chemicals and are apprehended by the same receptor machinery" (p. 12). There is no *substantial* difference between self and non-self and the "selfness and foreignness of an antigen depends on the interpretation given it by the immune system" (p. 12). No *essential* difference exists between self and non-self. The "origins" title suggests that experience is crucial for our ability to differentiate self from non-self. There are two sources of experience that help the immune system to differentiate between self and non-self: the genetic and the somatic. Evolution has endowed organisms with inherited mechanisms for handling infection through inflammation. Bacterial and viral products are identified by germline-encoded elements and objects identified in this context (*i.e.* antigens) are interpreted as non-self. In other words, it is the *context* of infection/inflammation that serves as the background for identifying foreignness. This idea has also been presented by Janeway (1992), who argues that the immune system evolved to discriminate the infectious non-self from the noninfectious self. This suggestion does not solve the conceptual difficulties associated with the concept of self, but does explain what context supports self and non-self discrimination.

Interpretation must assume not only basic familiarity with the "text" but *context* too. The somatic experience is the actual organization of the immune network in each individual. Somatic experience is no less important than the evolutionary one. We are all born with general templates for recognizing foreigners. However, actual experience is indispensable for the recognition of the threatening foreign.

The idea of somatic selection can be explained from an evolutionary perspective. It should be remembered (Langman and Cohn, 2000) that mammals, like human beings, have a relatively low rate of evolution (*e.g.* mutation) in comparison with bacterial and viral pathogens. Therefore, a germline selection might have been disastrous for them in any armed race with the pathogens. In other words, relying on genetic reshuffling of the antibodies would have been a poor evolutionary strategy. In contrast, somatic selection is better able to respond flexibly to the higher rates of mutation among possible pathogens.

"Harmony" is "the concern of the immune system: recognition of the right self-antigens and the right foreign antigens, interpretation of the context of recognition and a suitable response" (Cohen 1994, p. 16). It means that in contrast to the simple idea of self and non-self discrimination, the immune system is a highly orchestrated and contextual system of interpretation that transcends the simple dichotomy of self and non-self.

Surprising evidence supporting Cohen's thesis comes from the immunology of reproduction. Among the risk factors for SpAb is a form of testicular trauma. McLachlan (2002) suggests that "it is possible that even minor and/or repetitive sporting testicular trauma is sufficient" for the production of SpAb. This factor is explainable by Cohen's thesis. Testicular trauma, like a kick in the groin during a soccer tournament, may result in infection and inflammation in the damaged tissues. This context invites the identification of the sperm cells as foreigners and the production of SpAb for coping with them.

The idea of an infectious context has its critics. Anderson and Matzinger (2000) argue that the "infectious hypothesis" does not comport with the rejection of transplants by the host body. This rejection is observed when no infectious agents are evident. This critique is a serious challenge to the contextualist approach. The critique may be expanded to other questions. For example: Why does the immune system reject some tumors when a tumor is not accompanied by the context of infection? This is an open question that should be addressed within a contextualist theory of the immune self.

As a response to this challenge we may suggest that the immune system responds to the perturbation of regularity and that regularity is no more than embedded contexts of relations. For example, the reproductive system of mammals evolved in a way in which the fetus is developed in the uterus. This form of reproduction is regularity and, therefore, it is a context in which the fertilized zygote is tolerated. However, further theoretical elaborations will be later presented to cope with the difficulties of the contextualist approach.

Meanwhile, we should add another layer to our discussion by introducing the idea of the immune system as a complex system.

Efroni and Cohen (2002) argue that the immune system is a complex system that resists simple reductionism. Cohen locates his contextual perspective in the perspective of complex systems:

> The immune system is a paragon of complexity and needs the tools of complex systems research to understand it. (Efroni and Cohen, 2002, p. 24)

Indeed, achieving harmony is not a simple task. According to this suggestion, the observed properties of the immune system, such as self and non-self discrimination, are *emergent properties* that result from micro-level interactions between the heterogeneous constituents of the system.

The contrast between the complexity of the immune system and the simplicity of its function raises some questions. If the function of the immune system is as simple as suggested by Melvin Cohn and others, why should not we settle with a simple mechanism for explaining this function? Why complexity?

Henri Bergson (1911) discusses the contrast between the simplicity of function and the complexity of systems in his monumental work *Creative Evolution*. Bergson says that in analyzing the structure of an organ "we can go on decomposing for ever, although the function of the whole is a simple thing" (p. 94). This contrast should "open our eyes" since

> in general, when the same object appears in one aspect as simple and in another as infinitely complex the two aspects have by no means that same importance or rather the same *degree of reality*. In such cases, the simplicity belongs to the object itself, and the infinite complexity to the views we take in turning around it, to the symbols by which our senses or intellect represent it to us, or, more generally, to elements *of a different order*, with which we try to imitate it artificially ... (pp. 94–95; emphasis mine)

In other words, simplicity is an emergent property of complex micro-level interactions. To use an example from Bergson, a picture may present to us a simple figure: a sunflower by van Gogh, a dancer by Degas. However, when we try to understand these simple images by imitating them, we have to divide the picture into smaller and smaller pixels that together, on the macro-level, create the simple image. The immune system is exactly like this dynamic mosaic. It is composed by a variety of heterogeneous agents that cooperate and co-respond in a highly complex way, the same as an orchestra.

The macro-level product of this activity is the simplicity of the function: the emergent self and non-self.

We should be aware of the lesson taught by Bergson: simplicity and complexity belong to two different orders, or scales of analysis, and we should not confuse them by mistaking the simplicity of macro-level products with the complexity of micro-level interactions. We will further elaborate this issue but first another layer should be added to our understanding of the contextualist approach to immunology.

Cohen (Efroni and Cohen, 2003) does not consider the immune system as only a biodestructive system but as a regulatory system that is responsible for a certain portion of body maintenance. Wound healing, tissue repair, and cell regeneration are just some of the maintenance processes in which the immune system is involved. Rather than a warrior that defends his castle against invaders, the immune system is prosaically portrayed as the maintenance man of the apartment building we call the organism. This role is much less heroic but involves much more complexity. As one may know, the tasks of the housewife are much more diverse than those of the solider. There is no clear-cut, identifiable enemy or simple destructive activity. In this context, the simplicity of self and non-self discrimination is replaced by the complexity of meaning making. Antigens are identified not because they are signs of a non-self but because, in a certain context, certain biological agents do not integrate with the local maintenance activity of the organism and the meaning of this disharmony results in the immune response. This suggestion invites the question "what is the meaning of a context?" To answer this question we need to move to Valentine Volosinov and his contextual theory of meaning.

Cohen's contextualist approach in immunology clearly corresponds to the contextualist approach in semiotics. Let us dwell a little bit on this approach by using a wonderful example by Valentine Volosinov. Volonisov's work is a cornerstone in the meaning-making perspective that I develop in the book and one should pay close attention whenever his ideas are discussed. Consider the following scenario:

> A couple is sitting in a room. They are silent. One says, 'Well!' The other says nothing in reply. For us who were not present in the room at the time of the exchange, this 'conversation' is completely inexplicable. Taken in isolation the utterance 'well' is void and quite meaningless. Nevertheless the couple's peculiar exchange, consisting of only one word, though one to be sure which is expressively inflected, is full of meaning and significance and quite complete. (Volosinov, 1926, in Shukman, 1983, p. 10)

Understanding the sign "Well" in the above example, and our ability to extract the information it conveys is a *meaning-making* process that relies heavily on contextual cues and inferences. The meaning of "Well" is not encapsulated in the sign. The meaning is inferred by relying on contextual cues. What are these contextual cues? Volosinov suggests that we should examine the *non-verbal context,* which is formed from "(1) a *spatial purview* shared by the speakers (the totality of what is visible—the room, the window, and so on), *i.e.* the phenomenal field of the interlocutors; (2) the couple's *common knowledge and understanding of the circumstances—the result of years of being involved in patterns of interactions*—and finally (3) their *common evaluation* of these circumstances" (Volosinov, 1926, pp. 10–11, in Shukman, 1983), what Gregory Bateson describes as *belief.*

According to this suggestion, the sign "Well" is totally devoid of meaning in itself. If, however, we find that the two people are sitting in front of a window and see snow falling outside, and if it is winter where snow usually falls, the "Well" makes sense. *Meaning is therefore evident in our response to an indeterminate signal. Meaning cannot be determined in advance.* It is a response within a local context. A similar analogical thesis may be raised concerning self and non-self discrimination. The meaning of certain entities can be considered as self or as non-self only in context. The same agent may be ignored when it appears in the context of a healthy tissue and may be attacked in the context of a damaged tissue. Meaning, whether in semiotics or immunology, emerges in context. This idea will be elaborated further in the final section. However, before approaching the central issue, I would like to introduce the idea of a *relational approach of inquiry*.

It seems that one of the major difficulties in elucidating the meaning of the immune self is our essential approach to concepts. Sometimes we think like Platonists who seek the idea of the phenomenon under investigation. However, as I emphasize again and again, meaning cannot be found in the thing. There is nothing hidden beyond the curtain of the concept of the immune self. The immune self is a relational structure of interactions. This general approach to inquiry is presented and elaborated on in the next section.

7. The Nose and the Finger

Several years ago, I noticed my little daughter picking her nose. My wife demanded an immediate educational intervention to prevent a recurrence of this shameful activity by the *infant terrible*. As usual, her mistake was asking me to do this job. Instead of trying to convince the young toddler of the importance of this cultural norm, I challenged her older siblings with a

"Batesonian" (and, frankly, a misleading) question. "Hi", I said to them. "When Tamar picks her nose, which of them enjoys this activity more, her nose or her finger?" My older daughter, the first to reply, pointed to the nose as the source of the libidinal pleasure. Her brother, who is always happy to refute his sister's arguments, assumed the role of the anti-logos and argued that it was definitely the finger that enjoys the activity. "Both of you are wrong!" I declared in an authoritative manner. "The pleasure exists *in between*". My wife was shocked, the kids were amused, and my little daughter continued picking her nose. Indeed, in a culture in which objects precede relations, it is easier to explain pleasure in terms of objects (*e.g.* the nose, the finger, or the immune self) and their attributes than in terms of patterns of *relations* and *interactions*.

Bateson was one of the main figures who struggled to constitute an interactionist and contextual language of inquiry, which is highly relevant to our understanding of the immune self. In this section I introduce Bateson's ideas, in order to prepare the ground for a better understanding of the immune self.

In his seminal work *Mind and Nature* (1979), Bateson makes important distinctions among three terms: (1) *description*, (2) *tautology*, and (3) *explanation*. A pure description concerns the facts "immanent in the phenomena to be described" (Bateson, 1979, p. 81). A description contains information but no logic or explanation. In other words, it is a term that concerns the analytic list of components inherent in the phenomenon but without reference to the logical relationship among the components. It is just a set of differentiated components, a set of differences. A purely analytic mind may find the nose, the lips, the eyes, and the ears to be the components of a certain phenomenon. Nevertheless, without synthesis this list would never integrate into a whole—the face. The same difficulty might face the analytic immunologist who observes the multiplicity of immune agents without being able to integrate them into a working whole—a working network of immune agents.

In contrast to description, *tautology* offers connections between units of a description and contains no information. It is the logical infrastructure of the phenomenon and therefore corresponds to what we previously described as *the langue—the abstract system of signs*. Putting Mr. Potato Head's eye underneath his lips may turn the face into a monstrous nonsense image, one that, in terms of data, corresponds to a real face but lacks the internal logic that should organize it. This logic is always internal to the system under observation.

Bateson defines *explanation* as mapping description onto tautology. Explanation is the mental activity of mapping the micro-level elements onto an abstract macro structure, thereby giving the phenomenon meaning.

Bateson goes on to suggest that a process of inquiry, and let me add that a process of meaning making too, is a "zigzag ladder of dialectic between form and process" (Bateson, 1979, p. 191) and draws an analogy between form-tautology and process-description. As an illustration of this methodology he draws on his anthropological work in which he moved from a description of actions (a process) to a typology of sexes (a form), to interactions that determine typology (a process), to types of themes of interaction (a form), to interaction between themes (a process).

Bateson's conception amazingly resembles ideas presented by Bergson in his *Creative Evolution* (1911). Bergson distinguishes between the matter of our *knowledge* and its *form.* Matter is "what is given by the perceptive faculties taken in the elementary state". (p. 156). In other words, matter is the list of objects under inquiry. It is what Bateson describes as description. In contrast, form is the "totality of the relations set up between these materials in order to constitute a systematic knowledge" (p. 156). This is identical to Bateson's tautology.

Bergson further elaborates the distinction between form and matter with regard to the distinction between *instinct* and *intelligence.* He defines instinct as the utilization of a specific instrument for a specific object (Bergson, 1911, p. 148) or the faculty of using organized instruments. Instinct is a way in which we directly operate on the world without semiotic mediation. Through a direct, programmed, and germline-encoded operation of one object (or instrument) on another object.

Blinking as a response to a sudden approaching object is an instinct and engulfment of a bacterium by a macrophage is an instinct. It is a relatively stable and automatic use of the instrument (*e.g.* the macrophage). We cannot teach ourselves to avoid blinking without changing our structure. It is germline-encoded behavior. The innate immune system is an expression of an instinct. For example, phagocytosis of foreign particles is universal throughout the animal kingdom. Phagocytosis as a defense mechanism must involve a mechanism for allorecognition—the ability to recognize the difference between non-self and other individuals of the same species. Phagocyte recognition of foreign cells is achieved either directly by means of integral membrane recognition molecules or indirectly through soluble factors that bind to the foreign or to the damaged surface and mark it (Bayne, 1990). In both cases, recognition involves *pattern recognition* of a marker, which is a direct structural extension of the foreign or points directly at the foreign. That is, the innate immune system clearly operates as an instinct.

Intelligence is defined by Bergson (1911) as the "faculty of manufacturing artificial objects, especially tools to make tools, and of indefinitely varying the manufacture" (p. 146). Reviewing the seminal work of Luria and Vygotsky (1992): Signs are also tools. Therefore, intelligence is an open-ended activity of semiosis. It is not a simple expansion of an instinct by mediated knowledge but *a mediated process of semiosis.* The adaptive immune system expresses intelligent behavior as a somatically based system. It creates its own "tools" for handling pathogens. "This innate intelligence, although it is a faculty of knowing, knows no object in particular" (Bergson, 1911, p. 155). This statement does not mean that the immune system's ability to somatically respond is not genetically encoded. The adaptive immune system is an open-ended activity that is genetically embedded. In other words, the system's ability to transcend its own boundaries is encapsulated within the system itself, as implied by Bergson.

Bergson proceeds by suggesting that intelligence is the knowledge of form (*i.e.* tautology) and instinct the knowledge of matter (*i.e.* description). In other words, intelligence is knowledge of relations and instinct is knowledge of objects. The two forms of knowledge are inseparable and mutually dependent, as both Bergson and Bateson realized.

The implications of these ideas for the study of the immune system are clear. In this context, the phenomenon of the immune self may be described in terms of dialectic between form and process. It is the interplay of abstract logic, which organizes the fragmented experience of what we describe at a higher level of analysis as the "immune self", and the fragmented experience in itself. As Peirce and Bateson recognized, meaning demands a triadic relationship rather than a simple correspondence or a semiotic labyrinth, as suggested by the post-modernists' hall of mirrors or Jerne's network theory. In the next section I plan to lay out this methodology of inquiry, as well as describe the benefits that accrue to the study of the phenomenon of the immune self by analyzing the specific case of tolerance in the testes.

8. The Testes and the Immune Self

The production of sperm in to man's testes (or more accurately in his seminiferous tubules) is called spermatogenesis. Sperm are produced in the adult as the result of stimulation by anterior pituitary gonadotropic hormones. To review, hormones are chemical messengers that affect the behavior of cells in our body. They are signs that trigger specific responses in the body. These messengers regulate a variety of functions in a similar way to another important regulator, which is the nervous system. There are several hormones that play an important role in the generation of sperm. These hormones are: testosterone, follicle-stimulating hormone, estrogens, and growth hormone.

Sperm is created in several steps. The seminiferous tubules contain germinal epithelial cells called spermatogonia. A portion of these cells differentiate to form sperm cells. In the first phase of sperm genesis type A spermatogonia divide four times and form 16 more differentiated cells known as type B spermatogonia. The spermatogonia migrate among the *Sertoli cells*. These cells, which have been described by their discoverer as "nursing cells", have a cytoplasmic envelope that extends all the way to the central lumen of the tubule. The cells adhere to each other and prevent the penetration of immunoglobulin. In this mechanical sense the cells create a blood-tubule barrier that protects the evolving sperm cells. Figure 8.1 describes the sertoli cells.

The spermatogonia penetrate the barrier and become enveloped and protected by the Sertoli cells. For a period of 24 days, on average, the spermatogonium becomes a primary spermatocyte, and then divides into two secondary spermatocytes. Afterwards, the meiosis is slowly changed under the hospitality of the Sertoli cells into a spermatozoon (Guyton and Hall, 1996). In fact, the Sertoli cells play a crucial role in the production of

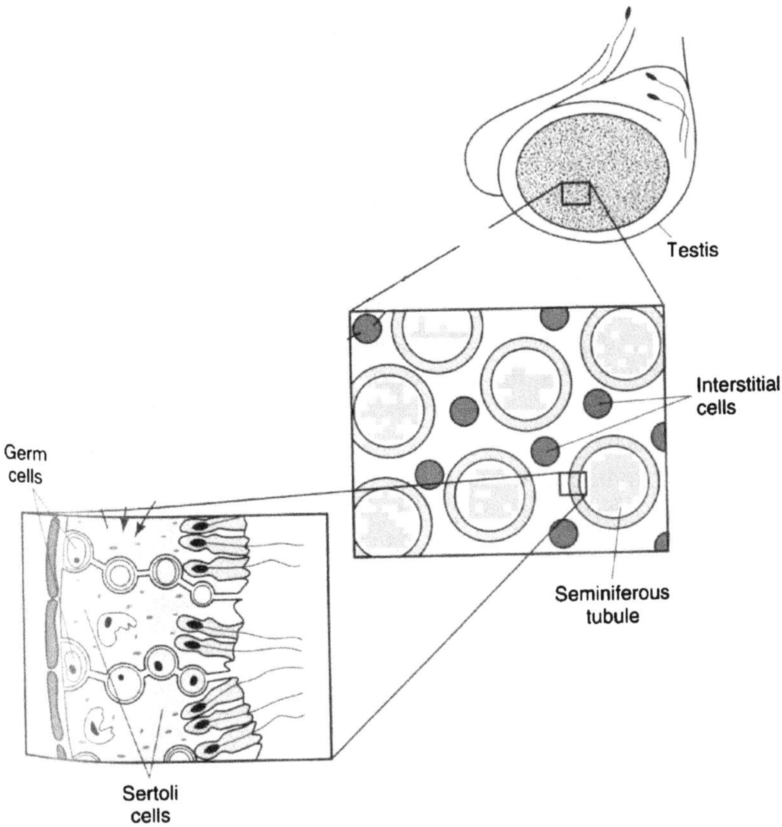

Fig. 8.1 A schematic representation of the Sertoli cells.

sperm cells and it is agreed that spermatogenesis in higher vertebrates is dependent on the function of Sertoli cells (Griswold, 1995).

The testes are an interesting site, from an immunological point-of-view. We know that when injected elsewhere in the body, sperm cells elicit a strong autoimmune reaction (Tung, 1980). In other words, these cells, which are a part of the biological self and necessary for its reproduction, are treated as non-self. However, these cells are secure in the testes, a phenomenon that led to the testes being considered as an immunologically privileged site like the brain.

The question, "why are there immunologically privileged sites?" is an interesting question in itself. Is there a common denominator between the brain and the testes? I am sure that from a feminist perspective some interesting and humorous analogies could be suggested, however, I do not want to get into this troubled water. My next step is to present the factors responsible for immune tolerance in the testes.

There are several factors that establish immunotolerance in the testes (Antonio Filippini *et al.*, 2001): the blood-tubular barrier, the local production of immunosuppressive molecules by Sertoli cells, and the Fas system, which regulates immunological homeostasis.

The blood barrier created by the Sertoli cells does not create a complete protection. In this context, it has been suggested (Antonio Filippini *et al.*, 2001) that immune tolerance in the testes is the synergetic result of two mechanisms: (1) "Physical" segregation of most of the autoantigens and (2) local production of immunosuppressive molecules by Sertoli cells (De Cesaris *et al.*, 1992). That is, the Sertoli cells secrete antilympoblastic proteins into the environment that suppress the activity of sperm antibodies. Now, we understand why in a case of infection or trauma to the tissue SpAb are released. The *context has been changed* and the suppression of lymphocytes is under suspension in order to temporarily allow them to contribute to the maintenance of the damaged tissue.

The contextual sensitivity of the Sertoli cells is important for understanding the immune tolerance in the testes. Sertoli cells can be used in the case of transplantation in order to avoid the rejection of the transplanted graft (Dufour *et al.*, 2004). That is, the tendency of the Sertoli cells to protect the evolving sperm cells is known and can be used in other contexts as well. However, this contextual sensitivity has negative implications too. If germ cells are forced to remain attached to the seminiferous epithelium for a period of time longer than necessary to complete their development, they will degenerate and eventually be phagocytosed by Sertoli cells (Mruk and Yan Cheng, 2004). In this sense, the Sertoli cells are not only the defenders of the sperm cells but in a changed context, their executioners.

Another interesting factor in immune suppression is Fas—a membrane protein that is a receptor for FasL, a cytokine belonging to the TNF family (Nagata, 1999). Fas/FasL interaction regulates the immune response by inducing apoptosis (regulated cell death) of lymphocytes. In mammalian cells, apoptosis may be triggered in two ways (Todaro *et al.*, 2004): extrinsic, which relies on a family of death receptors, and intrinsic, which is activated in response to cytotoxic stimuli such as DNA damage.

It has been argued that Sertoli cells expressing FasL interact with Fas-bearing autoreactive lymphocytes and destroy the lymphocytes through apoptosis (Bellgrau *et al.*, 1995). This suggestion has been controversial due to the simple fact that Sertoli cells do not directly interact with lymphocytes. However, Sertoli cells produce cytokines that communicate with agents of the immune system. Following this intricate web of interactions is extremely difficult and therefore Bellgrau's speculation cannot be totally dismissed.

Acknowledging the complexity of immune tolerance in the testes, Antonio Filippini *et al.* (2001) reject simple explanations of the testes being an immunologically privileged site by saying that "such simplistic interpretations are not plausible" (p. 447), and that "the mechanisms regulating the protection of the germline are necessarily complex, multiple and perhaps profitably redundant" (p. 447). Here we get into a general conclusion regarding the immune self. Self and non-self discrimination in the testes is a context-dependent and emergent phenomenon. There is no single mechanism that is, in itself, responsible for this distinction.

The components of a context suggested by Volosinov may be easily applied to the testes case. The *spatial purview* shared by the agents is the totality of the biological objects that exist in the local functional organ or complex. It is just as Bateson would have described it. Spermatozoa that appear in the testes appear in a spatial position that is immunologically legitimate. The *common knowledge and understanding of the circumstances—the result of years of being involved in patterns of interactions*—is the established pattern of relations between the objects. It is Bateson's tautology. It is the regularity, what Bergson described as the intelligence or the knowledge of the form. It is a common regularity among male mammals that sperm cells are created in the testes. Transferring sperm cells to another biological site would be a violation of this habit/regularity and would elicit a response. Finally, a "*common evaluation* of these circumstances" is produced by the immunological agents' complex process of communicating with and responding to each other. Hormones that signal the production of sperm cells and macrophages that sense the state of a tissue are just few of the agents that provide their input to the evaluation of the circumstances. In the case that one gets a kick in the groin, sperm antibodies might be produced because the evaluation of the circumstances, what I previously described as *hypothetical inference*, has been changed. The idea that the response of the system to a given entity is what defines the meaning of the entity is not new either in semiotics (Volosinov, 1986) or in immunology (Cohen, 2000a). A genuine contextualist always insists that meaning is not encapsulated in the message, which is in itself devoid of meaning, but in the process that results in a response to the message. In this context the immune system is not an exception, and immune tolerance in the testes is just a concrete example of this logic.

9. Conclusions

Tolerance and autoimmunity are usually associated with health and disease. However, as we have seen, autoimmunity may play an important role in protecting the body. Along the same lines, tolerance may play an important role in producing disease. As was recently argued:

> Recent evidence suggests that mechanism of tolerance normally exist to prevent autoimmune disease may also preclude the development of adequate antitumor response. (Mapara and Sykes, 2004, p. 1136)

This evidence suggests that the naïve value judgment put on autoimmunity and tolerance should be reexamined. Tolerance, as the absence of immune response to host constituents, might have in certain contexts deadly consequences. This lesson should direct us toward inquiring into the immune self as a theoretical construct which is the result of a meaning making process, a process in which an indeterminate signal such as a sign in natural language or a bacterium in the immune case is interpreted in context, resulting in a differentiated response that defines the boundaries of self and non-self.

If we adopt this perspective, then our self turns out to be a highly contextual and fuzzy concept that is *actively inferred* from raw data rather than *passively given* by our genes. This perspective can be illustrated through the case of malignant tumors.

Are cancerous cells a part of our self? In a case where a tumor development is associated with the acquisition of gene mutation and expression, immune recognition may get into action (Mapara and Sykes, 2004). In this case, the cells of a malignant tumor may be considered as non-self. However, in other cases the tumor's cells are not recognized because:

> Most antigens expressed by tumors are, in fact, normal self antigens to which deletional tolerance [tolerance through the elimination of the antigen-reactive cells] is likely to exist. (Mapara and Sykes, 2004, p. 1138)

So what is the general answer to the question, "Are cancer cells non-self?" The cells are the same cells in both cases. They are cancerous cells that are sometimes being tolerated and sometimes not. This fact cannot be changed, just the meaning associated with it. As we can see, there is no categorical answer to the question "What is the immune self?" Nor is there an algorithm to answer this question. The answer is to be looked for in the context, even the most trivial context of the spatial preview, as suggested by Volosinov. Indeed, it is argued that one of the factors that may determine tolerance of a tumor is whether it is localized in a place that is not accessible to circulating T cells (Mapara and Sykes, 2004). In this case, *being out-of-context is being meaningless*.

The idea of context-sensitive analysis as a paradigm for understanding cancer has already been presented, albeit in different words, by Vakkila and Lotze (2004), who argue that adult cancer might not be determined solely by cell growth, but also, by what we may think of as contextual factors such as sub-clinical inflammatory disease.

The inevitable question is how the immune system decides which agents to tolerate. The answer is that meaning is indeed evident *post hoc* and that context plays a crucial role in this decision. It is context that determines the meaning of the immune self. Context is an intricate and habitual network of agents that communicate and co-respond to each other. Following Volosinov, I have suggested that a context is composed from three major dimensions. The first dimension is the *spatial purview* shared by the agents. It is the totality of the biological objects that exist in the local functional organ or complex. According to this suggestion there is no single immune self just as there is no single meaning to a linguistic sign. The meaning of the immune self is determined by biological objects, which inhabit the local site, just as the meaning of a word is determined by its semantic surroundings. In the biological realm, knowing these objects is an instinct. It is the immune self's genetically determined knowledge about the local inhabitants. However, this knowledge is not only the knowledge of genetic markers; the picture is much more complex. It is possible that transplants are rejected not only because they bear the marker of non-self but because suppressive agents like Sertoli cells do not permit the tolerance of a tissue that does not emerge from the orchestrated, internal, and natural growth plan of the body. Our knowledge of biological development has not reached an appropriate level to scientifically consider this speculation, but it is still a speculation that warrants discussion.

Following the above discussion, we may conclude that the "immune self" corresponds to local responses of tolerance and attack, which result from the abstract and highly complex set of relations between the objects that constitute the functioning tissue. This is the *common knowledge*, what

Bergson described as intelligence: it is not the knowledge about objects that populate the local site but the knowledge about the relations between these constituents. For example, it is a common regularity among male mammals that sperm cells are produced in the testes and, in a matter of perfectly orchestrated timing, are protected by the Sertoli cells. Sperm cells do not bear the genetic marker of the immune self, which was established in the early development of the organism. They are not known through instinct but through a complementary form of intelligence. They are tolerated because they are anticipated as part of a natural process of sexual maturation among male mammals. Their arrival is announced by hormones, and within a local, spatial, and temporal window they are tolerated.

Finally, the "*common evaluation* of these circumstances", the third component of context, is a process of examining the correspondence between the objects in the situation (*e.g.* the presence of sperm cell) and the general pattern of relations that are supposed to organize these objects. It is a process of abduction or hypothetical inference as we discussed in the context of immune specificity. These relations are probably embedded in a genetic data that contribute, albeit in a complex way, the orchestration of biological processes. The programmed apoptosis of cells during the development of the fetus is an instance of this orchestrated relational construction. Following this line of reasoning, autoimmunity may serve as a normal process of maintenance as long as it is harmonized with context. On the other hand, autoimmunity may turn into a harmful activity if it deviates from this pattern of relations. To keep using the musical metaphor, autoimmunity turns into a disease when the orchestra fails to play in unison and harmony turns into cacophony. This metaphorical description is clearly evident in trying to establish tolerance in transplantation.

Transplants are rejected because they express tissue proteins that are identified as foreign to the host. Establishing tolerance to transplants is achieved by the long-term use of immunosuppressive drugs. Using immunosuppressive drugs as a strategy for producing tolerance seems the most natural move. As we may recall from the immune tolerance in the testes, this is one of the ways in which immune tolerance is actually achieved in the body. However, in a recent article, Waldman and Cobbold (2004) argue: "drug toxicity, chronic rejection, and immune deficiency (reflected in an enhanced incidence of cancer and infection) remain unresolved problems in the clinic" (p. 209). This suggests that a simple solution, such as the use of immunosuppressive drugs, is not the best answer to a complex problem. This confession concerning the limitations of immunosuppressive drugs teaches us a lesson about the highly contextual and orchestrated nature of immune tolerance; it may not be reduced to simple genetic markers. In the context of orchestration, cytokines play a crucial role. As secreted or membrane-bound proteins that regulate growth, differentiation, and activation of immune cells, cytokines may serve as a main communication channel for the evaluation of the circumstances. Indeed, evidence suggests that cytokines play a crucial role in cancer immunity and may be used for immunotherapy (Smyth *et al.*, 2004).

What are the implications of considering the immune self as a contextual construct? Identifying the objects involved in the immune response is a relatively easy task that has been conducted successfully. However, mapping the relations between this polyphony of agents is a demanding, integrated task. Understanding the correspondence between the objects involved in the immune response and the abstract, dynamic pattern of relations that organize their behavior is currently beyond our grasp. As the late Ray Paton (2002, p. 63) argued, "From a biological system's point-of-view there is a lack of tools of thought for dealing with integrative issues". However, we are currently in a better position to understand the immune self.

First, we understand that the meaning of the immune self, like the meaning of any other sign, is inferred from the response of the system to a given signal and it is not encapsulated in the signal itself. There is no positive definition of the immune self as suggested by the genetic-reductionist approach, there is no negative definition of the self as suggested by Burnets, and there is no post-modernist hall of mirrors in which the immune system is narcissistically occupied with itself. The immune self is defined *post hoc* as those objects to which the system responds with tolerance. It is defined through the *response* of the system as inferred from a contextual analysis.

The second implication of considering the immune self from a meaning-making perspective is that the immune self is flexible and tolerant enough to include symbiotic parasites. As argued by Lyn Margulis, symbiotic parasitism is the cornerstone of life forms on earth. For example, the mitochondria in our cells is hypothesized to be highly integrated and well organized former bacteria. Life forms emerged from cooperation no less than from competition, and the immune self, as a theoretical construct, should correspond with this wisdom. The above conceptualizes the immune self as tolerant of different constituents within the self. According to this suggestion, *E. coli* is a part of the self since the system's regulated response to this object is one of tolerance.

What is the general conclusion we may draw from the analysis so far? The conclusion is that the immune self is not a platonic, autonomous, and monolithic entity but a context-dependent construct. There is no self with a capital S. Some cells are considered as non-self due to the place and the timing of their appearance. In a constantly changing context they will be treated like a self. In other words, the question what is a non-self and self cannot be answered through a reference to a specific entity. Being a self and non-self depends on the response of the immune system in a given context, and this context is always a local context, as suggested by Volosinov.

Cat-logue 3

Bamba: Hi, Dr. N. What are you doing?

Dr. N: Well …

Bamba: The well is the same "Well" which is discussed in Volosinov's example? Full of meaning?

Dr. N: Well you did not give me a chance to finish my answer but well …

Bamba: Ok, I get your point. Let's move on to discuss the idea of context and reductionism.

Dr. N: Ok.

Bamba: It seems that the idea of context brings us to a dead end. If any sign is context-dependent then we can never say something general without it being qualified by the statement: "Yes, but it depends."

Dr. N: You are right. I'm afraid that my enthusiasm for context led me astray. We have a problem.

Bamba: Not if you shift the burden of proof to context.

Dr. N: What do you mean?

Bamba: Regularity and order should never be sought at the tokens' level. This is the wrong place. The reductionist immunologists looking for the meaning of the immune self at the level of the genetic marker committed exactly this type of mistake. Regularity exists in context and the number of contexts is significantly lower than the number of tokens that populate these contexts. The context of hierarchical relations is the same no matter who is the boss. There are different situations in hierarchical relations but for a human being to adapt to a social milieu it is important to ignore these particularities and to attune to the general scheme. Regularity exists only on the level of *habitual patterns of* relations. Therefore, meaning making can never be analyzed from a reductionist perspective.

Dr. N: Well, you are right.

Bamba: The same "Well," again?

Dr. N: You are so cynical! You definitely deserve a painful kick in the groin!

Bamba: My God! And to change the context of my beloved Sertoli cells?

Dr. N: Ah ha, so you do understand the meaning of context.

Bamba: You see, beyond humor and language games, we are both "materialists" that clearly believe that meaning, unless grounded, is meaningless.

Dr. N: Indeed. Meaning is grounded and our perspective is primarily grounded in our bodies. The perspective of a castrated cat would have been totally different from yours.

Bamba: Why go so far? Do you know Origen? He was a well-known personality from the early days of Christianity. One day he read the following passage attributed to Jesus by Matthew:

> For there are eunuchs who have been so from birth, and there are eunuchs who have been made eunuchs by men, and there are eunuchs who have made themselves eunuchs for the sake of the kingdom of heaven. He who is able to receive this, let him receive it.

Reflecting on this passage, and as a result of his religious zealousness, he castrated himself as an act of faith.

Dr. N: Ohh! This is definitely a changing context for the poor spermatozoa.

Bamba: Yes, for them it literally means to be out of context.

PART 2. WHAT IS THE MEANING OF THIS STORY?

From Syntax to Pragmatics

In the first part of the book, I criticized reductionism and pointed to its limitations. I argued that the world of living systems is much more complex than the reductionist rhetoric would have us believe and that we must seek an alternative conceptualization if we would like to better understand the realm of the living. The alternative I presented offers a biosemiotic perspective: a perspective that offers us an opportunity to examine the realm of the living as a realm that is constituted through semiosis (sign-mediated activity). More specifically, I showed that many biological phenomena that are currently conceptually obscure are clarified when considering them as activities of *meaning making*. The first part of the book illustrated meaning making in the context of the immune and genetic systems but the general idea of meaning making was only presented in a nutshell. The second part of the book aims to get us more deeply into the notion of meaning and into the meaning-making perspective.

I chose to locate this meaning-making perspective in the context of linguistics. This move is, to a certain extent, inevitable. As I previously suggested, the linguistic metaphor nurtured biological research, specifically that dealing with genetics. Therefore, it seems a natural move to employ the linguistic context for understanding meaning making among living systems. This move is comprehended only if we are familiar with relevant fields of linguistics. The following chapters aim to provide this background. However, a qualification should be added: I criticize the linguistic metaphor in biology and point to its shortcomings. I will argue that linguistic meaning is just a specific instance of meaning making in general and will delve more deeply into the meaning of meaning making. This section of the book is opened by a general discussion of meaning making in language and biology. The following chapters delve more deeply into syntax, semantics, pragmatics, and their relevance for understanding biological systems.

Chapter 9 Meaning Making in Language and Biology

1 Metaphorical Thinking
2 Living Systems and Boundary Conditions
3 Meaning Making
4 Meaning Making, Organization, and Disorganization
5 Meaning and Interaction
6 Implications
7 A Final Comment: "Let Truth Be Raised Up from the Ground!"

Chapter 10 God's Sacred Words

1 The "Book of Life" and the Book of Genesis

Chapter 11 It Means Nothing

1 Life is Meaning Making

Chapter 12 A Point for Thought: Meaning—Bridging the Gap between Physics and Semantics

Summary
1 Introduction
2 Information: Different Questions Lead to Different Answers
3 Information: A "Naturalistic" Perspective
4 Interactive Machines
5 Maxwell's Demon
6 Measurement as an Invention
7 Maxwell's Demon in a Klein Bottle

Chapter 13 The Rest is Silence

1 Wound Healing, Plasticity, and Context
2 Epigenesis
3 The Rest is Silence
4 How Do Biological Systems Know Themselves?

Chapter 9

Meaning Making in Language and Biology

1. Metaphorical Thinking

In the first part of the book I critically examined the limits of reductionism, adopted a semiotic perspective on biological processes, and pointed to the benefits of inquiring into various biological issues from a meaning-making perspective. The second part of the book is devoted to examining the issue of meaning making. Is it just an enlargement of the linguistic metaphor in biology? Or is a meaning-making perspective a more genuine venture?

Let me open the discussion by introducing the idea that human thinking is metaphorical in nature (Lakoff and Johnson, 1999). According to Lakoff and Johnson metaphors are not just rhetorical ornaments of our language, but an essential aspect of human language and thinking that emerged from basic bodily experience. For example, *knowledge* is considered to be an abstract concept. However, the most ancient metaphor of knowledge consumption is embedded in the physical organic act of incorporation through eating or sexual intercourse. Adam who knew his wife Eve in the sexual sense, Adam and Eve who ate from the tree of knowledge, and Jesus, who shared himself with his disciples through the symbolic act of the communion, are just some of the most famous examples in the Western tradition of knowledge consumption through physical (or organic) ingestion. In this sense, knowledge exists outside the subject as a concrete, physical substance and is literally incorporated through an organic process of assimilation.

Metaphors are the *sine qua non* of any process of understanding. Thus, a scientific theory must be critically examined for its reservoir of metaphors, and alternative metaphors sought that enlarge its scope and transcend its boundaries. Metaphors however should be critically studied. As Tauber (1996) argues:

> Theory must grope for its footing in common experience and language. By its very nature the metaphor evokes and suggests but cannot precisely detail the phenomenon in concern. (p. 18)

Indeed, metaphors are creatively generated rather than mechanically applied to a pre-given world (Shanon, 1992), and therefore they cannot detail the phenomenon of concern—an activity which is the role of the scientific model—but only guide the inquiry.

As discussed throughout this book, human language was metaphorically used for understanding biological processes, but is the linguistic metaphor in biology just a form of expression? For example, a popular book introducing genetics to the general public was entitled *The Language of the Genes* (Jones, 1995). However, Jones' book does not include in its index the terms *syntax* or *grammar*. In this case, the term *language* in the title is used as a form of expression and does not point to deep similarities between natural language and genetics.

Before delving deeply into the linguistic metaphor in biology, we should be familiar with an important distinction between the *representational* and the *non-representational* approaches to metaphor.

As Shanon argues, the discussion of metaphors in cognition and related disciplines assumes that metaphor is a relationship established between two *given entities whose attributes are defined prior to the establishment of their relationship*. This representational theory of metaphor is evident in Gentner's (1983) seminal work on analogy/metaphor as a form of structural mapping between two domains. For example, the analogy "an atom is like the solar system" is interpreted as a mapping of known, deep-structure similarities (*i.e.* similar relations) between one domain (*e.g.* the atom) and another (*e.g.* the solar system); electrons revolve around the nucleus just as the earth revolves around the sun. Although in some cases the use of metaphor may be interpreted by representational theory, Shanon propounds the alternative: that in other cases a metaphor has generative power to *create the similarities rather than simply assume them*. That is, the "metaphoric relationship is more basic than its constituents" (Shanon, 1992, p. 674), and the metaphor "creates new features and senses" (p. 674).

The linguistic metaphor in biology has mainly worked along the lines of the representational theory of metaphor and looked for similarities between human language and biological systems as two pre-given domains. The benefit of moving along this line of inquiry is questionable. If one is familiar with the pre-given properties of two domains, then finding similarities between them is of no use. Indeed, students of linguistics do not have to read *Essential Cell Biology* in order to understand human language, and students of medicine do not need to master Chomsky to understand cell biology. This critique of the representational theory of metaphor is not new to those familiar with theories of metaphor, and it casts serious doubt on the possible contribution of the linguistic metaphor to biology.

If we adopt the non-representational approach, our strategy should be different: First, we should draw the metaphor and only then examine the similarities that emerge from its use. This is the strategy that I adopt. To illustrate this strategy, let us examine the difficulties that result from using the representational approach to the linguistic metaphor.

Syntax is the study of regularities and constraints of word order and phrase structure (Manning and Schutze, 2003, p. 3). Syntax studies the grammar of language. The linguistic metaphor in biology has focused almost exclusively on similarities between the syntax of linguistic and biological systems (*e.g.* as evident in the structure of DNA). This is no surprise. Our knowledge of grammar has reached a high level of abstraction and formality that makes it easy to draw the analogy between the grammar of language and the "grammar" of DNA. However, as discussed throughout this book, the scope of linguistics is much broader than the study of grammar and in order to have a full grasp of a linguistic activity one must also study the *pragmatics* of language. To review, pragmatics is a field of linguistics that deals with language usage in context (Mey, 2001), in other words, the field of linguistics that deals with the generation of *meaning-in-context*. Although the generation of meaning-in-context is crucial for understanding biological systems, the linguistic metaphor in biology has, for the most part, ignored pragmatics. For example, Ji (1997), who propounds the idea of "cell language", describes human language as consisting of lexicon, grammar, phonetics/phonology, and semantics but ignores pragmatics. This ignorance may be explained by the tremendous difficulties facing pragmatics even in linguistics. However, this difficulty shows great promise for biology and linguistics/semiotics alike. The analogy between human language and biological systems may teach *both* biology and linguistics an important lesson on a difficult subject: how meaning emerges in context.

2. Living Systems and Boundary Conditions

Biological systems are open systems that exist on several distinct, complementary, and irreducible levels of *organization*. These levels constitute the systemic closure of the living system through feedback loops. In other words, they are *recursive-hierarchical systems*. It should be noted that this unique form of dynamic organization also characterizes the process of text reading and contrasts sharply with the organization of information-processing devices. Indeed, several scholars have argued that living systems are *reactive* rather than *transformatory* (information-processing) systems

(Cohen, 2000a). Transformational systems are sequential, linear systems that transform information in a specific order to achieve a specific goal (Cohen, 2000a). In contrast, reactive systems are multi-level, non-linear, ongoing systems that interact constantly with their internal and external environment to create sense out of their environment in an integrative gestalt manner that cannot be reduced to a digital binary code. In other words, living systems are meaning-making machines rather than information-processing devices: interactive machines (IM) rather than Turing machines.

Information, as classically defined by Shannon, is a probabilistic measure. As Emmeche and Hoffmeyer (1991) argue, unpredictable events are an essential part of life, and thus it is impossible to assign distinct probabilities to new events:

> The quantitative concept of information needs a closed possibility space. If the set of possibilities is open, one cannot ascribe precise probabilities to any single possibility and thus no information value. (Hoffmeyer and Emmeche, 1991, p. 3)

Their conclusion is that biological information must embrace the *semantic openness* that is evident, for example, in human communication, and that we should abandon the probabilistic conception of information. Indeed, the semantic openness of language allows the free interplay of ideas and concepts, just as a certain level of disorganization in living systems is necessary for the emergence of new forms. Without a basic level of disorganization, semantic openness cannot exist.

Following Bateson (1979), Hoffmeyer and Emmeche (1991) also propound the idea that living systems have two different codes: a digital binary code for memory (as in DNA) and a gestalt-type analogue code for behavior. I discussed this distinction in a previous chapter entitled: "Why Are Organisms Irreducible?" The syntactic approach to language emphasizes the digital aspect without paying attention to the analogue. However, if we want to understand living systems as meaning-making systems, then the analogue mode is indispensable.

The recursive-hierarchical and semantically open structure of living systems can be illustrated by protein conformation. As I previously discussed, a protein has an enormous number of potential conformations or organizations. Although a protein assumes an *energetically favorable structure*, we cannot understand its final conformation without taking into account several distinct and complementary levels of organization and boundary conditions imposed by the higher levels.

The final structure of the protein also depends on the context/environment: the interaction of the protein with a ligand. Therefore, in order to understand protein folding, we must take into account not only different levels of organization but also *interaction-in-context*. Metaphorically speaking, we must take into account the pragmatics of this process.

The idea that the living organism is composed of an irreducible structure was introduced by Michael Polanyi (1968), whose work I previously discussed. To review, one of Polanyi's main arguments is that an organism is a system whose structure serves as "a *boundary condition* harnessing the physical-chemical processes by which its organs perform their functions" (Polanyi, 1968, p. 1308; emphasis mine). In other words, "if the structure of living things is a set of boundary conditions, this structure is extraneous to the laws of physics and chemistry which the organism is harnessing" (Polanyi, 1968, p. 1309). If each level imposes a boundary on the operation of a lower level, then the higher level forms the *meaning* of the lower level (Polanyi, 1968), as evident in the folding of proteins.

Polanyi illustrates his thesis by turning to linguistics. According to Polanyi, the boundary conditions in living systems are analogous to the boundary conditions in linguistics. The meaning of a word is determined by the sentence in which it is located, and the meaning of a sentence is determined by the text in which it is located. This analogy has been previously discussed but should now be qualified.

First, although we cannot understand the words in a sentence without understanding the sentence, neither we can understand a sentence without understanding its words. This *hermeneutic circularity* was recognized long ago, and it seems to characterize the operation of living systems. However, due to a misunderstanding of the recursion process and the recursive-hierarchical organization of living systems, this hermeneutic circularity has been considered, at least by some philosophers like Russell, something to be avoided rather than a constitutive principle of living systems.

Second, biological systems may be metaphorically described as texts. However, there is no text without a reader. There is no meaning without an interaction. Therefore, both recursive-hierarchical organization and interaction are crucial for describing biological systems in linguistic terms.

The fact that Polanyi and others use the linguistic metaphor for understanding biological systems is not due to intellectual whim. Meaning making in natural language and the behavior of living systems do have something in common: Both take advantage of disorganization on the micro-level to create organization on the macro-level, through semiosis, recursive-hierarchy, and interaction. Both operate on the boundary of organization and disorganization to create meaning-in-context.

3. Meaning Making

Meaning making can be defined as a process that yields the system's differentiated response to an indeterminate signal. Later I will describe meaning making as sign-mediated interaction and meaning as the response or the effect produced by this interaction. For example, being an antigen is not an attribute that is explicitly or directly expressed by a molecule (*i.e.* the signal). The meaning of being an antigen is the result of a complex deliberation process (*i.e.* a meaning-making process) that is finally evident in the specific immune response (Cohen, 2000a). In this sense, meaning making is a process of computation in the classical etymological sense of assembling a whole from pieces.

The term *computation* is usually used in the technical, modern sense of a deductive process following a deterministic algorithmic program. However, Heinz von Foerster suggests restoring the original meaning of the concept. The word *computation* comes from the Latin *computare*, where *com* means together and *putare* means to contemplate or to consider (von Foerster and Poerksen, 2002). Meaning making involves bringing together different perspectives to achieve a specific response. This is what I describe as "symmetry restoration", which will later be discussed under the label: "transgradience". For example, the decision as to whether a specific agent is an antigen or not involves a variety of immune agents (macrophages, T cells, B cells, cytokines) that contemplate (*putare*) together (*com*) to yield the final immune response. In other words, the signal (*e.g.* an antigen) is *contextualized* in a wider network of immune agents to achieve a specific response. Meaning making is thus a process of computation in the analogue, holistic, and gestalt senses. To review, as a computation process it is vulnerable to loss of information and therefore meaning making, rather than simply generating information, actually involves the loss of information from one level of analysis in order to produce information at a higher level of analysis.

It should be noted that if there are no degrees of freedom in the system's response to a given signal, then by definition this system is not involved in meaning making. The potentiality of the signal[1] is a defining principle underlying communication processes in living systems. This flexibility may be illustrated through natural language, in which the same sign can be used in different contexts to express different things. The specific term for this phenomenon is *polysemy*. The benefit of polysemy is clear: "Polysemy … allows the use of the same word in different contexts and thus endows

[1] A signal turns into a sign whenever it is interpreted as signifying something.

language with indispensable flexibility" (Shanon, 1993, p. 45). This point is crucial for understanding both meaning making and the organization of living systems. In both cases, there is *maximum potentiality* (of the sign or the molecule) *at the micro-level that endows the system with tremendous flexibility for making sense* (linguistic or molecular) *on the macro-level.*

4. Meaning Making, Organization, and Disorganization

The term *sense* may have different meanings and connotations in biology and linguistics. Here, I use it as being closely associated with *organization.* Thus a signal (in biology) or a sign (in linguistics) makes sense if it is embedded in a higher-order structure of components (*i.e.* a context) that enables the system to produce a specific response. This point will be elaborated upon in the sections below.

The interesting thing about meaning making is its quasi-paradoxical nature and the fact that it operates on the *boundary of organization and disorganization.* Let me explain this argument in semiotic terms. The generation of meaning requires that the system has maximum freedom on the micro-level (*i.e.* the token level), but optimally minimal freedom on the macro-level. That is, *disorganization* in the sense of flexibility and dynamics is a necessary component of meaning making. In natural language, for example, we can produce an infinite number of meanings (*i.e.* responses, senses) with a limited number of words. Saussure pointed out that the meaning of a sign is determined by its location in a broader network of signs. In other words, the meaning of a sign is determined by different organizations of the signs among which it is contextualized. Thus, a "virgin" sign lacks any sense and can be linked to myriad organizations of signs; its degrees of freedom are potentially limitless. A sign is always a potential before it is mapped onto the macro-level of the sentence.

The above argument may be formulated in terms of Peirce's three ontological levels previously mentioned. The first mode, firstness, concerns pure potentiality. In the meaning-making process, it is associated with the most basic level of organization and with the potential variability of the signal. It is multiplicity of differences. It is a variety of differences and repetitions. To quote Peirce again:

> Freedom can only manifest itself in unlimited and uncontrolled variety and multiplicity; and thus the first becomes predominant in the ideas of *measureless variety and multiplicity.* (Peirce, 1955, p. 79; emphasis mine)

While the first mode (firstness) concerns pure potentiality, the second mode (secondness) deals with actualization of the potential through relations established between various components of the system. In this sense, secondness is a constraint—a form of organization—imposed on the first mode. In meaning making, the second mode of being is associated first with the grammar that constraints the possible meanings of the sign and second with a wider sense of contextualization in which a given sign is woven into the spatial purview, the common knowledge and the evaluation of the situation as experienced by the agents involved. In immunology, secondness, as a type of relationship, is evident when a specific signal is patterned (*i.e.* contextualized) into the web of immune agents and through this context obtains its meaning as an antigen.

The third mode of being (thirdness) is that "which is what it is by virtue of imparting a quality to reactions in the future" (Peirce, 1955, p. 91). In other words, it is the law, the habit that governs the behavior of the phenomenon, and our ability to predict its future behavior based on the law. It is the "conception of mediation, whereby a first and second are brought into relation" (CP 6:7). In meaning making, the third mode of being is evident when the system's different perspectives converge and are integrated to achieve a specific response in a given context, a process I describe following Bakhtin, as transgradience. It is a process of inference. In immunological recognition, as a process of meaning making, this mode of being is evident when the immune agents co-respond to each other through a complex communication network (Cohen, 2000a) to reach the final decision as to whether a molecule is an antigen.

5. Meaning and Interaction

The movement from disorder (firstness) to order (thirdness) is not random. This point can be illustrated through protein folding, an issue I previously discussed.

One might imagine that all protein molecules search through all possible conformations at random "until they are frozen at the lowest energy in the conformation of the native state" (Branden and Tooze, 1999, p. 91). However, this "random walk" would require far more than the actual folding time. In this sense, it is ridiculous to expect order on the macro scale of a living system to simply pop-up from firstness, just as it is ridiculous to expect a meaningful theory to emerge out of the uncontrolled delirium of a schizophrenic patient. In the case of proteins, it has been suggested that:

> The folding process must be directed in some way through a kinetic pathway of unstable *intermediates* to escape sampling a

large number of irrelevant conformations. (Branden and Tooze, 1999, p. 91; emphasis mine)

As Peirce argues, one cannot explain meaning by reducing it to lower levels of analysis; meaning is evident only once there is a triadic (or higher-order) relation between components, that is, only through the mediating force of thirdness. This bears repeating: meaning making happens only through sign-mediated interaction, through semiosis. In this sense, intermediates are needed to produce sense from senseless microelements.

The protein-folding process is a riddle. It is known, however, that the decrease in free energy is not linear. During the folding process, the protein proceeds from a high-energy, unfolded state (*i.e.* high potentiality) to a low-energy, native state through "*metastable* intermediate states with local low energy minima separated by unstable transition states of higher energy" (Branden and Tooze, 1999, p. 93; emphasis mine). To understand this process in semiotic terms, think about a word. It is transformed from a state of high potentiality to concrete actuality through the regulatory power of the higher linguistic levels of analysis (*e.g.* a sentence or paragraph). However, every move from one linguistic level of analysis to a higher level of analysis gives the word new potential for a different response. The word *love* has the potential to mean different things in different contexts. For example, when a man says to his wife: "I love you", the verb *love* is probably used in the sense of romantic affection although in order to determine the concrete meaning of the verb we should be immersed in the concrete context. In the sentence "I love books" the verb *love* is used in a different sense. When I declare, "I love books", I do not mean that I feel romantic affiliation with these objects or that I bring flowers to my books, or kiss them with passion. When a word such as *love* is in a sentence, its potential is actualized and the meaning of love is constrained. However, placing the sentence in a broader (extra-)linguistic context may result in a totally different response than the one expected from the sentence.

In the context of protein folding, this process suggests that in between the micro and the macro-levels of analysis there is a process of organization—one that has been described by Laughlin *et al.* (2000) as the "middle way" and by myself as the "*logic of in between*". This process seeks to overcome high-energy barriers to folding (*i.e.* constraints). Producing order from chaos is energy consuming, but without the system's basic tendency to reduce its free energy or to revert to a more basic mode of being, no work can be done and no meaning can be created. In this sense, the protein's natural entropic path to disorganization is subject to meta-regulatory processes (*i.e.* boundary conditions) that channel it so as to increase order. Like a Tai Chi master, the organism uses the natural tendency of its most dangerous opponents, the second law of thermodynamics or the natural tendency toward disorder, for its own benefit.

These metastable processes can be discussed in terms of interaction. The transfer of energy involves a weak coupling/interaction between at least two systems (*e.g.* ligand–receptor binding). This coupling is weak in the sense of the weak interactions discussed in research on synchronization. That is, one system/level of organization transfers energy to the other system/level of organization, but they both remain autonomous and separate systems/levels of organization. Do you remember that we previously mentioned the importance of weak forces in nature in general and in meaning making in particular? Weak interactions are a necessary condition for meaning making, whether in the biology (*e.g.* non-covalent forces) or in linguistics.

Returning to Peirce, we can understand that the pure potentiality of the micro-level is actualized by the repeated, habitual, or synchronized interaction of the third level. It is the third level that completes the triadic structure of meaning making. *Interaction is what mediates the emergence of meaning, whether in linguistics or in biology.*

Although meaning making assumes disorder, the specificity of response demands order on the *macro-level.* When a sign is located in a context, its degrees of freedom are significantly reduced. Thus, we need optimally minimal potentiality on the macro-level. For example, when using the sign *shoot*, we would like to have maximum freedom to use the same word once to express an order given to a soldier, and in a different context as a synonym for *speak*. However, in a given context, we want the sign to communicate only one of these meanings (*i.e.* to invite only one specific response) and not the other. To achieve this stability, the system has to be habituated through socialization in the case of human learning of signs, or through evolutionary processes (*i.e.* somatic learning) in the case of a biological response. In both cases, interaction plays a crucial role, as was previously discussed.

One must pay close attention to the fact that this habituation is not domestication. This is the reason why I used the expression "optimally minimal" to describe the decrease in potentiality. When we use language we would like to be understood through the social habituation of language practices. On the other hand, we always preserve the opportunity to be incomprehensible and vague through jokes, paradoxes, inventions, and other forms of nonsense that are crucial for a flexible life.

The need for macro stability is evident in protein conformation. A given sequence of amino acids forms a stable structure through the covalent forces that bind its molecules. However, this order is subject to non-covalent forces that interact to yield a specific conformation. Although a protein has an enormous number of potential conformations, it finally folds into one main conformation for the purpose of responding (for example) to a given ligand. In other words, the final conformation of the protein is determined through interaction with another biological entity (*i.e.* a ligand) in a given context (Cohen, 2000a). This description should be qualified, too. Although meaning making requires stability at the macro-level, it is not a total, rigid stability. The function of a protein requires structural flexibility and not rigid stability. The function of a word requires the same flexibility. You may want your word to achieve a specific response in a specific context, but you always want to reserve the ability to take your word back. Both in life and language, pragmatics assumes flexibility.

6. Implications

In this introductory chapter I have suggested that the linguistic metaphor in biology is based on a representational theory. I proposed investigating biological systems through another perspective to allow us to see the pragmatic, creative, and contextual side of language. If we are prepared to do this, then our next step is to discuss several aspects of biological meaning making along the lines we drew previously. First, we should emphasize the idea of biological *organization* rather than biological *order*. Whereas the term *order* usually pertains to information theory (*i.e.* the digital code) and the idea that a phenomenon can be represented (and quantified) through a unidimensional string of characters (*i.e.* the digital code), the term *organization* emphasizes the *multi-level structure* of biological systems and the interconnections among the components of the system. As Denbigh argues, wallpaper with a repeating pattern may be highly ordered but poorly organized (Denbigh, 1989). In contrast, a painting by Cezanne has a low level of order but a high level of organization that cannot be quantified (Denbigh, 1989). Language usage is organized rather than ordered and, as will be discussed later, it is dynamically organized. Any attempt to reduce meaning to order or pragmatics to grammar is doomed to failure. Following this line of reasoning, we may interpret the pathology of a biological system, as in the case of autoimmune diseases, as a problem of disorganization and meaning making—a problem with the system's ability to make sense out of signals by patterning them into a broader network of meaning. In this case, something happens in between the levels of organization; the boundary conditions do not function in such a way as to avoid the system's natural *entropic* reversion to firstness. Signs as functional generalities play an important role in this process. Without communication we are doomed to death.

As Harries-Jones (2006) argues, following Bateson, the death of a living system is more likely to be related to its loss of flexibility/resilience and to the devastation of its capacity to self-organize than to an outright loss of energy. The death of an organism might be the result of a vicious pathogen, but blaming the pathogen is of no help. This phenomenon should be investigated primarily through the failure of the immune system to make sense out of signals. For example, cytokines have been considered crucial to immune recognition. However, it was discovered that, under certain conditions, knocking out genes responsible for the production of cytokines did not destroy the immune system (Cohen, 2000a). This is not a surprising finding if one realizes that biological systems in general are characterized by overlapping and redundant feedback mechanisms. In this specific case, the immune system had organized *itself* to function properly. In other words, the immune system shows *resilience* in the sense that it renews itself and can therefore flexibly use other opportunities for meaning making and immune recognition. Later I will discuss life as meaning making and the ideas presented in the current chapter will be put in a larger context.

A mature understanding of a living system is possible only when the descriptions of the different levels are synthesized into a working whole. This is evident in immunology, where we have acquired knowledge about microelements and their interaction in the immune system but without an encompassing synthesis (Cohen, 2000a). For example, cytokines play an important role in communication among immune agents and there is a flood of information about cytokines. However, it was argued that practically nothing is known about the behavior of the (cytokine) network as a whole (Callard *et al.*, 1999). This problem, which was mentioned by Callard and his colleagues, years ago, should not be underestimated. Unless several distinct but complementary levels of organization are integrated and it is shown how they influence each other, the behavior of the immune system is to a large extent incomprehensible. This conclusion is, in fact, an invitation for researchers to investigate the recursive-hierarchical structure of living systems and the unique way these systems make sense out of their environment.

7. A Final Comment: "Let Truth Be Raised Up from the Ground!"

Ferdinand de Saussure made the famous distinction between *langue* and *parole*. To review, whereas the former concerns language as an abstract system, the latter concerns the concrete and contextual use of language (*i.e.* speech). According to Valentine Volosinov (1986), whose work has been previously mentioned, this dichotomy is false and should be transcended. However, the representational theory accepted this dichotomy and directed the linguistic metaphor in biology to examine the *langue* of biological systems rather than their *parole*: the syntax rather than the pragmatics.

Is there an alternative? A Talmudic story about the creation of man gives us a hint.

In *Genesis Rabbah* (1985, pp. 78–79), an ancient Jewish commentary on the Book of Genesis, we find the following story concerning the creation of man. Let me present and explain the story piece by piece.

> Said R. [Rabbi] Simon, "When the Holy One, blessed be he, came to create the first man, the ministering angels formed parties and sects".

In the first excerpt we are informed of the context (*i.e.* the creation of man) and told that the ministering angels were divided with regard to whether God should create man.

> Some of them said, "Let him be created", and some of them said, "Let him not be created".
>
> Mercy said, "Let him be created, for he will perform acts of mercy".
>
> Truth said, "Let him not be created, for he is a complete fake".
>
> Righteousness said, "Let him be created, for he will perform acts of righteousness".
>
> Peace said, "Let him not be created, for he is one mass of contention".

The division is between the angels that represent Mercy, Truth, Righteousness, and Peace. While Mercy and Righteousness support the creation of man, Truth and Peace oppose it. God, as a democratic governor, seems to notice that the vote is evenly split. For good reasons half of his ministering angels support the creation and for good reasons the other half oppose it. How can a decision be made in this difficult condition of equality? This question also bothered the Talmudic sages, who ask:

> What then did the Holy One, blessed be he, do?

The answer they provide is surprising, since God's action appears to be a form of madness:

> He [God] took truth and threw it to the ground.

The ministering angels seem to be shocked by this "crazy" move because Truth is the property most identified with God. Therefore they ask God to explain his behavior:

> The ministering angels then said before the Holy One, blessed be he, "Master of the ages, how can you disgrace your seal [which is truth]?"

The response they receive from God [or the angels?] is interesting and demands interpretation:

> Let truth be raised up from the ground!

A possible interpretation of God's answer is that, in the real world, creation (of man, meaning, or life) based on ultimate truth is not possible. Ultimate truth is an impossible foundation for creation and action, just as the abstract structure of language cannot, in practice, guide the generation of meaning in context.

We also learn this lesson from Swift's *Gulliver's Travels* (Swift, 1994). In *Gulliver's Travels* an architect proposes a new building method—from the roof down to the foundations. This method has its rationale, since it protects the builders from the rain and the sun! Indeed, by building from the' "roof" down we are protected from reality. The linguist who studies the abstract systems of *langue* is protected from the "disturbing noise" of the *parole*, just as the builder is protected by the imaginary roof from the disturbances of sun and rain. However, in neither case can language be understood or a house built. The Talmud does not dismiss the importance of the abstract. Nevertheless, the abstract and the ultimate, rather than guiding creation, should grow organically from creation or, as the Talmud suggests, "be raised up from the ground". This Talmudic stance turns on its head the common Western-Platonic conception in which the abstract and the ultimate come before the material and the concrete. According to the Talmudic story, desire (organic growth, the truth that rose from the ground), and madness/arbitrariness (God who makes a decision by throwing truth on the ground) are essential for understanding the genesis of meaning. In this context, the logic of in between (sense/nonsense, abstract/concrete, and method/madness) may be the appropriate logic for guiding our inquiry into meaning making in both language and biology.

The next chapters present different domains of linguistics and language activity and push the idea that, in both life and biology, pragmatics has precedence over grammar.

Chapter 10

God's Sacred Words

Those of us who are blessed with children have been amazed again and again when our young baby starts using language. From a misunderstood babbling emerges a unique, powerful, and mysterious form of representing the world, communicating about the world, and at the highest level, creating worlds. Indeed, the ability to use language has been considered by many scholars as the property that differentiates human from non-human organisms. According to this conception, the words of Ecclesiastes (3:19): "Man has no pre-eminence above a beast" should be replaced by "Man has pre-eminence above a beast as a speaking creature".

Great advancement has been gained in understanding human language, and in this part of the book I would like to briefly introduce the reader to several key ideas in linguistics. First, it is important to understand that spoken language is a multifaceted process, which is studied by those in different sub-disciplines of linguistics and associated fields such as psychology, neurology, philosophy, and computer science. Studying acoustics and the articulation of speech is the mandate of *phonology*. The most basic voices we produce are organized into meaningful units called *morphemes*. For example, the word *dancer* is composed from two morphemes: *dance* and the suffix *er*. Studying morphemes is the mandate of another field of linguistics—*morphology*. Our language is built in hierarchical order. The voices we produce are organized into morphemes, and morphemes are organized into words, organized into phrases, and sentences. *Syntax is the field of linguistics that deals with the structure of sentences.* In other words, syntax is the branch of linguistics that studies how the words of language can be combined to "make larger units like phrases and sentences" (Baker, 2003, p. 265). Translating these sentences or the words from which they are composed into thoughts and ideas is an activity inquired by *semantics*, and the way language is actually used in interaction, in context, is studied by *pragmatics*. In a nutshell this is the scope of modern linguistic analysis.[1]

[1] I do not discuss the whole spectrum of modern linguistic research and exclude from my analysis interesting fields such as computational linguistics.

My discussion of linguistic topics is guided by its relevance for understanding the biological realm. Therefore, phonology and morphology will not be discussed and we will start directly from syntax that studies the structure of sentences. As any other dynamic field syntax has different schools and varieties but there is no doubt that just as the field of psychoanalysis originated from Freud's theory, the modern field of syntax originated from the theory of Noam Chomsky. Chomsky's main contribution is in identifying the linguistic universal in the field of syntax (Sampson, 1980). Although there are many languages in the world, Chomsky argued that there are underlying syntactic structures that characterize human language. Chomsky was mainly occupied by the rules that govern the organization of words/phrases in a sentence. These rules actually generate the sentence and this is the reason for using the term *generative grammar*.

Two things should be noted. First, Chomsky's theory focuses on *descriptive rules* (Carnie, 2002)—the rules that actually guide the way people construct sentences. In other words, the rules that Chomsky was seeking were not stylistic rules taught by teachers but the universal rules that govern human language. In this sense, Chomsky's thesis is clearly a scientific thesis that aims to transcend the particularities of a phenomenon (*i.e.* a given language) in order to uncover the generalities that govern nature, in this case human nature.

Syntax is actually interested in whether a sentence is properly put together and not whether the sentence is meaningful (Baker, 2003). The rules that govern the structure of a sentence are called *grammar*. Grammar helps us to understand, or more accurately define the difference between a well-formed sentence and a sentence that is not well formed. For example, consider the following two sentences:

1. Danny saw himself in the mirror.
2. Themselves saw Danny in the mirror.

The second sentence is ill formed, but why? The explanation concerns a unique family of nouns that end with -self like *himself* or *herself*. This family is known as *anaphor*, and English grammar forces us to have an antecedent to anaphor and that an anaphor agrees in gender and number with the designated subject. When I wrote the second sentence using Word, the computer informed me that this sentence is not well formed and advised me to revise its grammar. Indeed, the second sentence violates the rules of using an anaphor. The anaphor *themselves* lacks an antecedent and does not cohere with the gender of Danny, who is probably a male, and with the number of individuals (in this case only one) who are involved in the situation.

In what sense is the above scenario represents a "rule"? It is clear that this grammatical rule is not the same as a rule in the natural sciences. In the natural sciences a rule is an abstract description of an observed regularity exhibited by nature. For example, the first law of thermodynamics states that energy can be converted from one form to another, but it cannot be created or destroyed. When I kick a ball some of the energy that is stored in my muscles turns into a form of kinetic energy, which moves the ball. Some of this energy is converted into the kinetic energy of the ball and some of the energy dissipates into the environment in the form of heat. An inventor who would like to create a new source of energy is not really a creator. He or she is actually inventing new ways of using existing energy, but energy is not created or destroyed. The first law of thermodynamics cannot be violated and as the common expression suggests: You cannot beat City Hall.

The fact that a grammatical rule can be violated indicates that it is a rule in a different sense. Is it a habit, a psychological rule? And what does *psychological rule* mean? Is there a universal psychology of the mind? The answer to the first question is positive. A grammatical rule is a psychological rule. Chomsky's rules adhere to the psychological realm. These rules of grammar can be violated, and sound awkward to the native speaker, but they try to formalize the regularity that characterizes a given language. It is the human judgment that determines whether a sentence is well formed, or whether it is ill formed and violates a certain rule of grammar. In practice, there are serious difficulties in defining a well-formed sentence. In any case, no linguistic regularity exists outside the human mind. This conclusion immediately raises the question: Is there, in biology, an analogue to the idea of a well-formed sentence? We will get to this point later.

A sentence is composed of several *syntactic categories*, such as noun, verb, preposition, and adverb/adjective. The reader may question the term *syntactic categories* for describing categories that can be allegedly defined semantically. For example, to decide whether a certain word is a verb, we should decide whether the meaning of this word represents some kind of action, or state of being. However, the attempt to categorize the meaning of words from a semantic perspective is problematic due to the polysemy of signs. For example, the same word can be classified into two different categories based on its structural positioning in the sentence. Consider the following sentences:

1. "Sinners which *knowledge* their sins" (Tyndale[2]).
2. "Why have I found grace in thine eyes, that thou shouldst take *knowledge* of me?" (*Ruth* 2:10).

[2] English religious reformer and martyr (1494–1536).

In the first sentence *knowledge* is used as a verb and in the second as a noun. The above example illustrates the difficulty of determining a part of speech by using semantic criteria. This fact brings us to the idea of a layered system that was discussed in one of the previous chapters. The structure of the sentence is composed from words and we cannot determine the meaning of the words by using the words themselves. The meaning of a part is not encapsulated within the part. The meaning of a part can be determined only in relation to other components of the analyzed whole. In other words, *meaning is relational and it emerges from the unique positioning of a component within a whole.*

In linguistics, we determine the category in which a part belongs by examining where the words appear in the sentence and what kinds of suffixes they take. For example, a noun is the object or the subject of the sentence, it follows determiners such as *the*, it is modified by adjectives, and so on. These rules may help us to determine the category to which a word belongs even if we are not familiar with the meaning of the word. For example, read the following sentence:

> The Kaghoratik ate its breakfast.

I am sure that you are not familiar with the meaning of the word *Kaghoratik*. It may raise some speculations (sounds like a Russian name) or associations but the word in itself is not familiar to you. It is a word I have invented just this moment. Nevertheless, you can easily determine, by using the hints presented previously, that *Kaghoratik* is a noun.

Why is it important to determine the category of a word? Why should we care whether it is a verb or a noun? The answer is that grammar is an essential layer in understanding the meaning of a word, and the organization of words may help us to understand their meaning. Consider the following sentence:

> Can you ____ me?

The empty space can be replaced by different words but not by all the words in the lexicon. The unique position of the empty space signals to us that the missing word is a *verb*. If we have to search a huge database of words in order to find the appropriate one for this empty slot, then the constraints provided by the structure are very helpful in reducing the amount of work we should invest in the search. The rules of syntax or more accurately *regulatory of* constraints *in words appearance* may reduce the uncertainty associated with the category of a given word. This is a powerful device in the acquisition and use of language. A mother may turn to her son and say, "Look! This is a

Chambalala". The child may not be familiar with the word *Chambalala* but he may correctly infer that this is a noun, an object to which his mother points. The grammar supports the *meaning* of words by reducing the uncertainty of its tagging. Reducing the uncertainty is exactly the defining characteristic of information. In other words, regularity at the syntax level is highly informative. Unfortunately, Chomsky produced a sharp divide between syntax and semantics. Consider the following sentences (Chomsky, 1975):

1. Colorless green ideas sleep furiously.
2. Furiously sleep ideas green colourless.

The two sentences are not normally produced in English and therefore, from a statistical perspective, they are both remote from English. Nevertheless sentence (1) is non-sensical but grammatically correct, while sentence (2) is both non-sensical and grammatically incorrect.

This example shows that a sentence can be non-sensical and grammatically correct but also that the information value of two sentences with different "grammatical information" can be the same. Chomsky argued that any statistical model based on the frequencies of word sequences would assign equal zero probability to the two sentences. This statement can be questioned. As suggested by Zellig Harris (1991):

> Given the great number and the changeability of the sentences of language, the word combinations cannot be listed. However, they can be characterized by listing constraints on combination, such as can be understood to preclude all the non-occurring combinations, leaving those which are indeed found. (p. 4)

Harris is not a single voice in this debate and other leading linguists such as Michael Hoey (2005) challenge Chomsky's thesis and present the idea of linguistic regularities of sentence components in terms of constraints' satisfaction and information value. Harris' is also a radical statement about the relation between syntax and meaning. While Chomsky put a sharp demarcating line between semantics and syntax, Harris points to the way in which constraints support the emergence of meaning. For example, in the sentence "Sheep eat grass" we take the verb *eat* to be an operator on the argument pair sheep/grass. In the context of this specific argument, it is more likely to find the operator *eat* than to find the operator *drink*. After all, sheep eat grass but they do not drink it. The partial meaning of *grass* as matter, rather than as liquid, is determined by the operator *eat*. To quote Harris (1991) again:

> The meaning of words are distinguished, and in part determined, by what words are their more likely operators or arguments; and

> the meaning of a particular occurrence of a word is determined by the selection of what words are its operator or argument in that sentence. (p. 5)

This statement echoes Saussure's seminal teaching that the meaning of a sign is always negatively created within a semiotic network and that: "Language is a system of interdependent terms in which the value of each term results solely from the simultaneous presence of the others" (Saussure, 1959, p. 114). Although, Harris does not mention Saussure, his argument concerning the emergence of meaning out of constraints clearly adheres to Saussure's seminal teaching.

Harris' focus on constraints, and the embedded nature of linguistic constraints, is in line with information in its naturalistic sense and the thesis presented in this book. A crucial aspect of meaning making, whether in linguistics or biology, involves constraints that are imposed on a lower-level mold. As illustrated in the Book of Genesis, creation is primarily the constraint of chaos. To conclude we may argue that:

> *Meaning is created when a given system's degrees of freedom are constrained in a regular way for a given observer, that is another system, which uses the regularity as an input for its maintenance and functioning.*

This statement emphasizes the fact that a living system is characterized by having constraints as an inherent part of the system. In other words, and following Polanyi, we may say that life's irreducible structure is grounded in boundaries that are inherent to the system. This point makes a sharp division between a living and a non-living system.

It is important to understand that organisms have in-built constraints. At the beginning of the first chapter of *Developmental Biology* (2000) Gilbert says something interesting about the difference between a machine and a living organism:

> One of the critical differences between you and a machine is that a machine is never required to function until after it is built. Every animal has to function as *it builds itself*. (Gilbert, 2000, p. 3; emphasis mine)

In other words, living systems are autopoietic systems (Maturana and Varela, 1992) that compute themselves. They are not simply executable computer programs, and in this context the question, "Who reads the book of life?" becomes more pressing than ever. If living systems are self-constructed

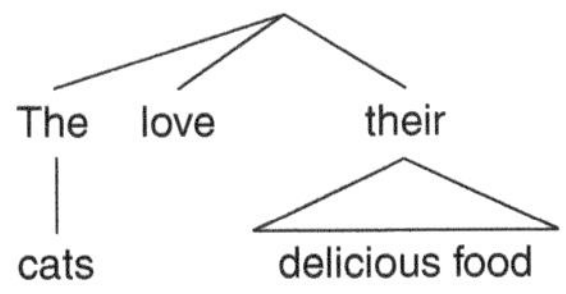

Fig. 10.1 A syntax tree of the sentence.

systems, then meaning cannot simply be reduced to their genetic strings, even as a metaphor for understanding living systems. Chomsky's machine-style grammar is irrelevant. In this context, the organization of the micro-level components is just one of many layers of constraints that should be taken into account to serve the task of interpretation. Interpretation is inevitable in both natural language and genetics.

To better understand the nature of linguistic constraints we should turn to *constituents*. A group of words that functions together as a unit is called a constituent.[3] A constituent is an important notion in syntactic theory. Moreover, constituents are embedded one inside another and create a *hierarchical structure*. Figure 10.1 shows a sentence and a graphical tree representation of its hierarchical structure.

Notice that the tree structure of the sentence is actually a *graph*. A whole theory of graphs was developed in mathematics, and the representation of a sentence in a graph makes it a possible candidate for a semi-mathematical analysis. Mathematics deals with the abstract structure and relations between certain objects. In our case, we may want to inquire into the abstract structure of sentences and therefore our analysis is mathematical in nature.

Indeed, mathematics is clearly evident in Chomsky's thinking. In his *Syntactic Structures*, Chomsky (1957) suggested that grammar is defined by a finite set of initial strings and a finite set of instruction formulas of the form $X \rightarrow Y$ interpreted as "rewrite X as Y". For example, a sentence is a string that can be rewritten as composed from a noun phrase and a verb phrase. Formally:

$$\text{Sentence} \rightarrow \text{NP} + \text{VP}$$

[3] A formal definition of a constituent is that a constituent is a set of nodes exhaustively dominated by a single node. The term *exhaustive domination* is defined as follows: Node A exhaustively dominates a set of nodes B, C, ..., D, provided it immediately dominates all the members of the set and there is no node G immediately dominated by A that is not a member of the set.

Indeed, "normal" sentences always include a subject NP and a verb and the subject appears before the verb. Given such grammar, the structure of each sentence can be represented by a hierarchical tree diagram such as the one above. Later, in his *Aspects of the Theory of Syntax*, Chomsky (1965) showed how a structural description might be generated by a system of explicit rules of transformation. In this sense, Chomsky's generative grammar is a powerful descriptive tool for linguists.

Let us turn from linguistics to biology. Chomsky made an important contribution to linguistics. However, it is not clear what contribution he has made to biology. Adopting the linguistic metaphor in biology inevitably raises this question. Is Chomsky's theory of syntax a relevant metaphor or analogy for biology? Analogy is defined as similarity in some respects between things that are otherwise dissimilar. Is there a similarity between the grammar of human language, the rules of syntax, and the "grammar" of genetics? Are the strings of amino acids governed by the same rules as those governing the grammar of natural language? No one will accept this analogy and it is meaningless. Maybe, the grammar of human language may be metaphorically used as a rhetorical ornament for biologists? So, is the syntactic approach relevant to genetics? For good reason, the ideas of Chomsky seem to be of minor relevance for biologists. Let me support this statement. I searched the databases of several leading journals in bioinformatics and examined the extent to which the name of Noam Chomsky, the world's most cited living intellectual, is mentioned in the journal's papers. I started with *Bioinformatics*, the leading journal in the field of bioinformatics. The name of Chomsky, and therefore the reference to his work, was mentioned only in eight papers. In *Journal of Computational Biology* the name Chomsky was mentioned only in four papers that were published between 1999 and 2005. I was quite desperate and went to *Nature Genetics* to see whether the name Chomsky has any relevance to those who are directly occupied with the sequencing and the study of the DNA. No results were found.

Although the attempt to use linguistic methods for biological research was considered to have promise (*e.g.* Searls, 2002; Dong and Searls, 1994), we may conclude by saying that the linguistic metaphor in biology in its purely syntactic sense is more a metaphor in the rhetorical ornamental sense than a metaphor that exposes deep similarities between human language and biological systems. I believe that the power of the linguistic metaphor for biology is in raising our awareness about the constraints imposed on tokens in a string of letters as a part of a meaning-making process. However, focusing on the string level and excluding the rest is a wrong move, as will be illustrated in the next section.

1. The "Book of Life" and the Book of Genesis

In 1994, a strange paper was published in the journal *Statistical Science* (Witztum *et al.*, 1994). This paper, written by Witztum and his colleagues, has the rather technical title: "Equidistant Letter Sequences in the Book of Genesis". The paper was the basis for a bestseller. It was reprinted in full in the bestseller and therefore it is possibly the "most printed scientific paper of all time" (McKay *et al.*, 1999).

This paper was a most pretentious attempt to look for hidden patterns in a religious text and its final conclusion is that the Book of Genesis includes hidden messages. Who is the author of these hidden messages? The ultimate and surprising answer is "God" himself.

One may immediately dismiss the apparent obscurity of this paper. However, the paper passed an extremely long process of peer review that took five years. Moreover, only five years after the publication of the paper a critical and scholarly response was published in the same scientific journal.

In his introduction to the critical response to the paper, the editor Robert E. Kass (1999) argues that when reviewing Witztum's paper the reviewers and some members of the editorial board were not "convinced that the authors had found something genuinely amazing" (p. 149). He argues further that the journal published this paper

> in the hope that someone would be motivated to devote substantial energy to figuring out what was going on and that the discipline of statistics would be advanced through the identification of *subtle problems* that can be arise in this kind of pattern recognition. (Kass, 1999, p. 149; emphasis mine)

Prof. Kass sounds as if he is making an excuse rather than offering an explanation. In his editorial introduction to the paper that was written five years earlier:

> Our referees were baffled: their prior beliefs made them think the Book of Genesis could not possibly contain meaningful references to modern day individuals, yet when the authors carried out additional analyses and checks the effect persisted. The paper is thus offered to Statistical Science readers as a challenging puzzle. (Kass, 1994, p. 306)

This is an interesting statement. First of all, it does not present the referees as open-minded scientists but as secular zealots with a strong predisposition against Witztum and his colleagues' religious beliefs. Prof. Kass could have chosen different words. He could have argued that the reviewers do not believe that the Book of Genesis contains statistically established patterns of words. In this case, the Book of Genesis may not contain statistically significant patterns of words although it may still have "meaningful references" in other senses to many individuals. It should be noted that some of the individuals who found meaning in the Bible were not ignorant peasants but the most prominent scientists of their time.

Although the "mindful experts" of *Statistical Science* were inclined against the paper, they were unable to point to its fallacies. If these fallacies were as "subtle" as Prof. Kass argued in his 1999 editorial introduction, then how is it that the expert referees and the members of the editorial board were unable to find them? If the paper included "subtle problems" why was so much energy needed in order to detect them? I believe that the answer to these questions is that the "mindful" experts of the journal were unable to grasp the gap between statistical patterns and their meaning. They confused regularities with meaning. Because regularities were established through accepted statistical tools they inferred that they were meaningful. However, since their predispositions led them to expect no meaningful patterns in the text, they were baffled.

The extraction of meaning out of a sequence is not a simple task of identifying statistical regularities but a process of *interpretation*. This is the lesson we may want to learn from this stimulating case. Therefore, the aim of this section is to describe and analyze the case of the Biblical code and to examine its relevance for the extraction of meaning out of biological texts.

Witztum's paper is based on the concept of equidistant letter sequences (ELSs). This simply means that we select letters in a given text by using an equally spaced distance between the letters. For example, the following sequence is a 1-D array of letters.

YGHAKLIEMRSKNWSESGUCVMSLAAQN

This sequence looks quite meaningless. It has a start, which is the letter Y, it has a certain length but it looks like an arbitrary collection of letters. Let us apply the ELS methodology and extract every fourth letter beginning from the first letter on the left. Let us define *d* as the skip between the letters. If we apply this method we extract the letters that compose my name:

YGH**A**KL**I**EM**R**SK**N**WS**E**SG**U**CV**M**SL**A**AQ**N**

Let me calm the excited reader. This is not God's message but a pattern I encrypted in the sequence.

T	S	H	T	I	M	S
E	I	D	S	R	A	B
G	U	R	S	E	M	A
T	L	B	F	O	K	O
K	O	K	G	F	B	N
Z	A	O	R	P	Q	V
L	K	S	H	D	F	J

Fig. 10.2 A matrix with an encrypted message.

The encrypted message does not have to be limited to a single-dimension array. An encrypted message can appear in a text written as a 2-D array. For example, see Fig. 10.2.

This is a 7×7 2-D array of letters. Use a skip of $d=2$ starting at the upper left letter and you will be exposed to God's own opinion of my book!

We may consider a text as a 2-D array, and the letters as points that appear on a straight line. To avoid the problem of the line ending at the vertical edge of the text and reappearing at the opposite vertical edge we may "paste" the two vertical edges of the text. The result is a cylinder in which the sequence of letters spirals down, like a snake that never ends. In fact, the name of this snake in mysticism is the *Uroboros*. The Uroboros is a snake that gives birth to itself through its own mouth and by that creates the appearance of a never-ending and self-penetrating being. My colleague Steven M. Rosen (2004) and I have both used this symbol in our books to illustrate the idea of a re-entering form.

Witztum *et al.* (1994) organized the Book of Genesis in this way. They found that ELS uncover words with related meaning in close proximity. The fact that we can find words in a given text by using the ELS methodology is not so very impressive. In an article published in *Skeptical Inquirer*, David E. Thomas (1997) writes:

> "Hidden messages" can be found anywhere, provided the seeker is willing and able to harvest the immense field of possibilities. But do they *mean* anything? (p. 36)
>
> Thomas answers this question with a definite "No!"
>
> The promoters of hidden-message claims say, "How could such amazing coincidences be the product of random chances?" I think the real question should be, "How could such coincidence not be the inevitable product of a huge sequence of trails on a large, essentially random database?" (p. 36)

Thomas is attributing the emerging patterns to fallacious methodology. However, things are not so simple. Witztum *et al.* (1994) admit that:

> ELS's for short words, like those for פטיש (the Hebrew word for 'hammer') and סדן (The Hebrew word for 'anvil'), may be expected on general probability grounds to appear close to each other quite often, in any text. (p. 430)

However, they argue that in Genesis, the phenomenon persists when *d* is confined to a minimal skip. Their methodology was as follows. They used a sample of pairs of related words and examined the significance of their close proximity. Moreover, they developed a statistical model that expresses how close "on the whole" are the words making up the sample pairs.

The sample of words was built from a list of rabbis and their dates of birth or death. It was found that the overall proximity of personality and his date was far from the one expected by chance. The authors' conclusion is that "the proximity of the ELS's with related meaning in the Book of Genesis is not due to chance" (Witztum *et al.*, 1994, p. 434). Halleluiah!

The question whether the sacred text hides meaning is relevant for studying whether a biological text hides meaning. One may easily dismiss this analogy on the ground that while the meaning attributed to the Book of Genesis was intentionally encrypted in the text by a specific sender (*i.e.* God) to a given receiver (*i.e.* the reader), there is no sender in a biological text. In other words, no superhuman deity is responsible for sending us the information in the DNA string. Well, a prominent theoretical biologist by the name of John Maynard Smith has a different point of view. In a paper examining the concept of information in biology (Maynard Smith, 2000), he argued that natural selection is the coder of meaning into the string of the DNA. The deity behind the Book of Genesis was replaced by the deity called the blind watchmaker. Halleluiah again!

What is the problem in trying to uncover hidden patterns in a sequence or more exactly in extracting meaning from grammar *per se*? There are different problems, of course, ranging from methodological statistical issues such as the appropriate benchmark for comparison to interpretation of regularities. For example, sequence analysis in bioinformatics suggests that if two sequences share statistically significant sequence similarity, they will share significant structural similarity. One should not confuse similarity and homology. Any two sequences have some quantitative measurable similarity. However, homology implies that "the similarity has some special meaning, specifically common ancestry" (Pearson and Wood, 2001, p. 40) and therefore homology is a *qualitative* rather than a *quantitative* measure. Homology is inferred from sequence comparison. If we have two sequences then we can hypothesize that the two sequences were created by chance or that they have a common ancestor. If we estimate the probability of the first hypothesis as significantly low then it may be refuted and the other hypothesis may be accepted. This dogmatic methodology in bioinformatics is problematic. The researcher is playing against chance. However, chance is not an explanation but our way of conceptualizing our ignorance. In addition, this process includes human decisions, which are beyond the sterile appearance of statistical sequence alignment: for example, choice of sequence segment to be aligned, choice of algorithm, choice of weights assigned to matching/mismatching residues, and so on (Cvrckova and Markos, 2005).

The interpretative stance is indispensable for meaning making. Witztum and his colleagues are wrong to try to reduce meaning to grammar (more accurately to linguistic co-locations) just as reductionist biologists are wrong to try to explain a biological function through a genetic sequence. Those who try to deny the significance of interpretation are fundamentalists who seek God's intentions in encrypted messages. This criticism holds for fundamentalists whoever they are whether religious fundamentalists who seek God in the statistics of the Biblical text, or scientific fundamentalists who seek their own God by analyzing sequences of DNA. In this context, another lesson should be learned from Gregory Bateson. Envisioning the linguistic metaphor in biology, Bateson (2000) argued that both

> grammar and biological structures are products of communicational and organizational process. The anatomy of the plant is a complex transform of the genotypic instructions. And the "language" of the genes, like any other language, must of necessity have contextual structure. (p. 154)

In other words, the digital code of DNA and its organization is meaningless unless it is interpreted in context. The context is the one that restrains (to use Bateson's terminology) or constrains the entropy of the system and its natural tendency toward disorder. Context is

> the collective term for all those events which tell the organism among what *set* of alternatives he must make his next choice. (Bateson, 2000, p. 289)

Bateson (2000) further argued that emphasis on a contextually sensitive analysis is a watershed differentiating reductionism and cybernetics, which he considered an alternative framework for understanding living systems:

> It may (perhaps) be true in physics that the explanation of the macroscopic is to be sought in the microscopic. The opposite is usually true in cybernetic: without context, there is no communication. (p. 408)

That is, contextualization is a top-down process. It is the macro-level organization that constrains micro-level elements. This observation suggests that contexts are embedded and can by themselves be messages for a higher-level context. "Contextual structures could themselves be messages". Organisms operate by interpreting signs in given contexts. Sometimes the context turns to be the message rather than the background for the interpretation of a message. Therefore the context is a "*meta*message which *classifies* the elementary signal" and a context of a context is "a meta-metamessage which classifies the metamessage" (Bateson, 2000, p. 289).

No context, no meaning. No inference, no meaning. The rules that constrain the organization of micro-level particles are just one aspect of meaning making and one should be careful not to assign them a hegemonic role in the multifaceted process of meaning making. Grammar can never be a substitute for meaning but just one layer of meaning making. God's intentions, and I doubt whether this expression is of any sense, cannot be found in grammar. In contrast with the famous expression "God is in the details", it seems that God uses details only as a platform. The next chapter aims to take us a step forward in the voyage toward meaning and to present Valentine Volosinov's unique and thought-provoking theory of meaning. This theory that states that meaning means nothing will necessarily push us forward to a pragmatic perspective of language.

Chapter 11

It Means Nothing

In the previous chapter, I introduced the field of syntax and presented the rather trivial argument that meaning cannot be reduced to grammar. If meaning cannot be reduced to grammar then maybe the relevant place to seek for meaning is in the correspondence between a word and its conceptual counterpart. Semantics may seem to be the next relevant stop on our journey. Semantics is a sub-discipline of linguistics that deals with meaning, albeit in a very unique sense. Although semantics has different roots and schools, a commonly held position today is that semantics is a "mentalistic enterprise" (Jackendoff, 2003, p. 267) meaning that "we are ultimately interested not in the question: What is meaning? But rather: What makes things meaningful *to people*?" (Jackendoff, 2003, p. 268). This postulate clearly restricts the study of meaning to human beings. The cell actively interprets signals. However, it does not have a "language of thought" in the sense of mental representations *à la* Fodor. Thus, if the process of semiosis takes place among other creatures and even at the cellular level, then semantics does not have anything to offer to us! This disappointing conclusion should not lead us to despair. Meaning is not the intellectual property of a discipline or a given school of thought and we may seek for answers in different places.

Clarifying the meaning of meaning is the aim of this chapter. As one can easily realize clarifying the meaning of a concept is not an easy task especially if one is trying to recursively understand the meaning of meaning. Indeed, meaning is understood in a multitude of ways and the interested reader is invited to consult dictionaries for the multitude of senses associated with the word *meaning*. The fact that *meaning* can be considered in a multitude of ways should not be considered as a problem but as a defining characteristic of meaning. Meaning is basically about the multitude of ways in which a differentiated signal (or a string of signals) may be responded to. By accepting this suggestion, we turn the problem (of meaning having different senses) into a defining characteristic of meaning. Generally speaking, meaning concerns the way in which an indeterminate signal (*i.e.* a signal that can be responded to in a multitude of ways) is constrained, contextualized, and responded to. A cell with an antibody is involved in a process of meaning making because it responds to signals (*i.e.* molecules) that bind to the antibody. The cell responds to this signal in multitude of ways. Sometimes the response will be ignorance and sometimes the response will be the initiation of a biological cascade. When encountering a suspicious molecule, the B cell is in the same position as we are when we encounter the sign *meaning*. Both the cell and the person have the potential to respond to the signal in a multitude of ways according to a given set of contextual cues and inferences. I have no intention to deny the differences between a cell and a human being; however, the preliminary and basic similarity should be acknowledged.

My attempt to encounter the issue of meaning making in biological systems is somehow unconventional since it has been extensively influenced by an unconventional source—Valentine Volosinov. Volosinov was an intellectual from Bakhtin's circle who published his seminal work *Marxism and the Philosophy of Language* during the late 1920s. Reading this manuscript, one immediately notices that the title of the book is an attempt to pay lip service to the Marxist regime rather than to incorporate Marxism into Volosinov's linguistic theory. This lip service did not help, and it seems that Volosinov, who disappeared in the late 30s, paid with his life to the regime he was trying to satisfy.

Volosinov's book has great relevance for understanding the meaning of meaning and I dwelt on this issue in my previous book. Here, I would like to simply introduce the theory and show its relevance for understanding the meaning of meaning.

Volosinov's starting point is that language should not be considered as an abstract system but as a mode of communication through signs. In this context, the basic linguist unit of analysis is not the sign but the *utterance*. An utterance is different from a sentence. A sentence can be repeated. However, an utterance is not repeatable but always different depending on who says it and under what conditions (Mey, 2001, p. 199). This statement clearly blurs the boundary between syntax, semantics, and pragmatics, and emphasizes linguistic communication as a pragmatic event. In this sense, a sentence is a decontextualized and atemporal abstraction of an utterance—a singularity of a communication event. For example, the sentence "I love you" has been repeated many times; too many times in modern history by individuals in real-life, in songs, and in movies. We can analyze this sentence syntactically but this analysis will bring us to nowhere. When considered as an utterance, "I love you" is a singular event in which a particular individual addresses another particular individual in a concrete and unrepeatable context of interaction. Without understanding the particularity of the utterance we will be in the position of an "armchair linguist". From a scientific perspective this position is untenable. One cannot accept the idea of a physicist who can formulate Newton's laws but have no ability to calculate the trajectory of a falling body. In the same vein, one cannot accept the idea of a linguist who can formulate general rules of linguistic organization while having no ability to comment on the meaning of a live utterance. In this sense, an utterance is the basic unit of Parole (dynamic linguistic activity), and it should be the focus of our study. Traditionally, linguists have focused on the sentence as their unit of analysis. As shown in the previous chapter this focus is specifically true for a grammatical analysis. However, as argued by Mey (1998):

> The supremacy of the sentence by no means has been recognized by all linguists. In particular, the Russian school of linguistic thinking, as embodied in the names of Volosinov and Bakhtin ... has been a forceful defendant of alternative ways of considering the sentence. (p. 93)

Volosinov's insistence on the *utterance* as the basic linguistic analysis is not a whim but a calculated move that aims to emphasize the *interactive (dialogic) and context-dependent* nature of any communicative act. This nature is evident in human and molecular interaction alike. When ligand–receptor interaction takes place it is always a singular event in which a decision has to be made whether the ligand is an antigen or not. *From now on whether I speak about a sign, a sentence, or a molecule as a part of a communication activity, I discuss them as utterances.*

The focus on the singularity of the communicative act and the attempt to theorize it has an in-built problem. One must be aware of *the paradoxical nature of trying to say something general about uniqueness.* This paradox is also evident in the work of Bakhtin. As argued by Holquist (1990b):

> It is precisely the radical *specificity of individual humans* that he [Bakhtin] is after: a major paradox in all Bakhtin's work is that he continually seeks to generalize about uniqueness. (p. xx)

Volosinov and Bakhtin emphasize the uniqueness of the utterance. However, as modern thinkers I believe that we should not be threatened by the attempt to discuss the generalization of uniqueness. After all *singularity* is a scientific concept.

Let us summarize. The basic unit of linguistics is, according to Volosinov, the utterance, which is *the basic and unrepeatable unit of communication between interacting agents.* This unit can be a sound, a sign, or an argument. The boundaries of the utterance are not determined *a priori* by structural linguistic features but by its communicative function. Therefore, the meaning of the utterance is a property belonging to the utterance as a whole. Volosinov describes the significance of the utterance as its *theme* and emphasizes again that the theme is determined not only by the linguistic forms that comprise it but also by extra-verbal factors of the situation (*i.e.* context). For example, the utterance "I love you" can be used by Romeo to address his beloved Juliet, but also by a glutton chef who observes his new dish and communicates to himself his own self-satisfaction. Again, the idea is that the internal linguistic form of the utterance is not enough for understanding its meaning. Today, this is the common wisdom of those who try to develop computational tools for language comprehension. Context is crucial for understanding the meaning of an utterance and context is always local and concrete. Because Volosinov uses theme in the sense we use for meaning, from now on I will use the term *meaning* in the Volosovian sense as the significance of the utterance.

Volosinov uses the term *meaning*, as distinguished from *theme*, to denote the aspects of the utterance that are reproducible and self-identical in all instances of repetition. In other words, meaning is

> the technical apparatus for the implementation of theme. Of course, no absolute mechanistic boundary can be drawn between theme and meaning. (Volosinov, 1986, p. 100)

Meaning, the technical apparatus for the implementation of theme, involves the digital code and it needs the analogical code as a complementary facet for meaning making. Volosinov is actually differentiating between the relatively stable sense of the utterance (*i.e.* its syntactic "deep structure") and its dynamic and context-dependent aspect. He realizes that both stability and dynamics are necessary for communication and points to the delicate balance between the two and on their mutual interdependence. For example, the basic sense of the sign *chicken* denotes the domestic fowl bred for flesh or eggs. This is the denotation of chicken; its meaning. However, let us imagine that a group of teenagers are examining their courage by jumping from a tree. One of the teenagers is standing on the branch looking down with anxiety. Paul, the leader of the group, is encouraging him by shouting: "Come on Jack, don't be a chicken!" In this specific context, the theme, the concrete meaning of the sign *chicken*, is not the same as the denotation of chicken. The meaning/denotation of chicken is used in the concrete metaphorical sense of coward. "Come on Jack, don't be a coward!" The theme of the utterance relies on the denotation of chicken. Without the generally agreed upon denotation of chicken it is impossible to make concrete use of it in the given context.

The idea that the constituents of an utterance (signs) have a stable base and a dynamic potentiality brings us to the polysemous nature of signs (or words). It has been realized by Volosinov that "multiplicity of meanings is the constitutive feature of words" (Volosinov, 1986, p. 101). This statement can be generalized by saying that multiplicity of meanings is a constitutive feature of the *sign*. Signs, whether in human language or in the immune system, have the potential to be responded to in different manners. In a case, where there is only one response to a sign then it is not a sign but a signal. As argued by Volosinov (1986):

> Meaning, in essence, means nothing; it only possesses the potentiality—the possibility of having meaning within a concrete theme. (p. 101)

Theme is an attribute of the whole utterance. If we would like to examine the concrete meaning of a sign then it can be done only in the context of the whole utterance. Let me illustrate this point by using an example taken from Chapter 3 of *Alice's Adventures in Wonderland* where the following dialogue takes place:

> 'Edwin and Morcar, the earls of Mercia and Northumbria, declared for him; and even Stigand, the patriotic archbishop of Canterbury, found it advisable –'
>
> 'Found *what*?' said the Duck.
>
> 'Found it', the Mouse replied rather crossly: 'of course you know what "it" means'.
>
> 'I know what "it" means well enough, when *I* find a thing' Said the Duck: 'It's generally a frog or a worm. The question is, what did the archbishop find?'
>
> The Mouse did not notice this question but hurriedly went on …

This amusing dialogue illustrates the polysemous nature of signs by taking the indexical *it* as a concrete example. The indexical *it* points to some object. However, the exact nature of the object cannot be inferred from the indexical itself or from its relative positioning in the sentence. The Duck correctly realizes that for it/him, *it* usually signifies a frog or a worm. This is the meaning of it. However, the Duck also realizes that the content of the indexical *it*, that usually has the value frog or worm, may have different senses to different observers.

After understanding the crucial difference between meaning and theme we should move forward. Volosinov emphasizes the communicative nature of the utterance. However, the process of communication he is discussing is far from the Shannon model some of us have in our minds. A communication process, according to Volosinov, is a process in which all the parties take an *active* role. In this context, to understand another person's utterance means "to orient oneself with respect to it" (Volosinov, 1986, p. 102). In other words, "Any true understanding is dialogic in nature" (p. 102). In this sense, meaning is the "effect of interaction between speaker and listener produced via the material of a particular sound complex" (pp. 102–103). This statement is extremely important and can be extended far beyond human communication to biological systems:

> *Meaning is the effect that is produced via semiosis through interaction between at least two parties.*

This is the definition of meaning I adopt at this stage. Meaning is the significance of the utterance and the significance is the effect of the mediated interaction. According to Volosinov's definition, meaning is a functional term. It is functional in the causal sense. Meaning is the result of interaction and it is the effect of a certain semiotically mediated interaction on something. There is no meaning without a mediated interaction and there is no point of speaking about meaning without pointing at the effect of this interaction. Here we are getting to Volosinov's final point, which concerns the evaluative aspect of the utterance. Volosinov suggests that every utterance is above all, an *evaluative orientation*. An utterance is not a communicative picture of the world as suggested by the early philosophical writings of Wittgenstein, for example. Language is *orientational* rather than denotational (Becker, 2000). It is a way in which we attune ourselves to context, to the particularities of the here and now, through semiotic mediation. It is coordination (with others) under constraints (context). The idea that meaning is orientational emphasizes the active nature of meaning making. Meaning is always actively implied rather than passively given. "To imply means to fold something into something else (from the Latin verb *plicare,* to fold)" (Mey, 2001, p. 45). While using language-in-context things unfold and meaning is implied. The creation of meaning is characterized by implicature. Here is an example. Sometimes when asked by my young daughter "What time is it?" I answer by saying "Time to sleep". My answer does not directly address the question but that it is understood is implied by my daughter's laugh. She knows that it is late and that by answering "Time to sleep" I am not pointing at a specific hour of the day known as "Time to sleep". Instead, my answer has the illocutionary force of ordering my daughter to go to bed, and she understands the meaning of my utterance by fully grasping the contextual cues and the implication derived from my utterance. Her laugh indicates that she has actively extracted the meaning of my utterance and has oriented herself toward a given psycho-cultural context in which a father is ironically responding to his young daughter's question. In sum, meaning, rather than a simple homocentric correspondence between a sign and a concept, utterance, or idea, can be considered as an emerging product of interaction that results through semiotic mediation from the active concerted efforts of communicating agents to attune them to context. The implications of this idea are enormous. What is life if not meaning making?

1. Life is Meaning Making

In her beautiful novel *The Stories of Eva Luna*, Isabel Allende (1992) tells the story of a young woman by the name of Belisa Crepusculario. This woman was born to a family so poor that they did not even have *names* to give their children. Reading this poetic description of ultimate poverty, we as academics should bless our good fortune. We always have names to give things, and at least in this sense we are extremely wealthy.

As academics, we should not only enjoy the wealth of names we have for describing the world but also ask ourselves what is the *meaning* of the names, concepts, or signs that we use so freely. Life itself is a concept that deserves such an inquiry. Biologists study living systems, but if you ask a biologist what life is the answer you would probably get is less than satisfactory. The biologist may smile and say that although he is dealing with organisms, the concept of life is beyond his direct and specific field of expertise. He may send you to the department of philosophy or worse, to some of his colleagues who have philosophical pretensions.

The problem of clarifying the meaning of the major organizing concepts of a field is not unique to biology. Richard Feynman once commented on the difficulties of understanding the meaning of energy. In fields where even a formal definition is not at hand, the situation might be worse. Ask a psychologist, "What is the 'self'?" and you will find yourself spinning into a labyrinth of words.

It is known that the definition of life is far from satisfactory. Although *pre-theoretically* (*i.e.* intuitively) the concept of life seems to be the most comprehensible of all, and although we have a wealth of knowledge concerning some of the mechanisms underlying what we pre-theoretically conceive as living systems, we actually fail to successfully define life and to draw the boundary between living and non-living systems.

Identifying the characteristics of living systems, or what Koshland (2002) describes as "The Seven Pillars of Life", is not enough. By itself, no characteristic of a living entity seems to be sufficient for a definition of life. On the other hand, a definition of life that includes *all* of the characteristics seems to be too narrow. For example, the Darwinian evolutionary aspect of life does not allow us to include life at the level of a single object as *life-here-and-now* (Luigi-Luisi, 1998). In addition, technological advances add another layer of complexity to this problematic state of affairs. Today, we ask ourselves not only whether viruses are living entities, but also whether computer viruses are alive.

As Cleland and Chyba (2002) suggest, "the philosophical question of the definition of 'life' has increasing practical importance" (p. 387). This point was re-emphasized in a paper by Conrad and Nealson (2001) published in *Astrobiology*. These authors point to the need to define life in the most general, universal, non-earthly sense in order to support the goal of comprehensive detection of life with appropriate conceptual tools. In this section, I would like to respond to this challenge by approaching life as a *meaning-making process*.

I qualify my response in one major respect: I do not pretend to offer a new philosophers' stone that will turn problematic definitions of life into scientific gold. Rather than offering a new definition, I offer a new perspective and ask pragmatically what *work* (Lange, 1996) this new perspective may provide to the study of life. As I have shown throughout this book, a meaning-making perspective may offer a new way for thinking on a variety of phenomena from junk DNA to immunity in the testes. I hope to show that the idea of life as meaning making follows the same successful path.

Let me begin with a simple observation and a statement we encountered before. What differentiates a living entity from a non-living entity is that a living entity can respond actively to its environment (internal or external) by turning a difference into a "difference that makes a difference" (Bateson, 2000). In other words, a living system is a system that actively turns differences into informational content, which is used for the autopoiesis of the organism. This process is mediated of course through the use of signs as I have suggested in Chapter 4: "Why Are Organisms Irreducible?"

In contrast with a living system, a non-living entity has, to borrow the old phenomenological expression, an "existence in and for itself". In this sense, a non-living entity is *indifferent* to its environment. It is not that a non-living matter is non-responsive to the environment. As time unfolds rust appears on my old bicycle and clefts appear on my old wooden chair. When I throw an elastic rubber ball against the wall it loses, then regains its original shape. However, these differences are structural differences in response to a direct and unmediated encounter with physical forces. These changes are not the result of a difference that makes a difference.

Following Bateson, we may suggest that life in its most general sense is machinery that actively turns fluctuations that can be described in physical/chemical terms into behavior that can be described on a *different logical level of analysis* in biological or psychological terms (Conrad and Nealson, 2001). Like

Bateson, I believe that this shift in languages, or logical levels of analysis, from the physical/chemical to the biological is indicative of a living system that exists on several logically distinct but complementary levels of organization.

For several reasons, Bateson's idea of a "difference that makes a difference" may be an appropriate theoretical springboard for discussing the concept of life. For example, it encapsulates the possibility (but not the necessity) of Darwinian evolution but does not preclude the possibility of non-Darwinian evolution. It is also relevant for describing life as *life as it is here and now* (cellular) but also as *life as it initially was* (transient forms that preceded cellular life), and *life as it could be* (artificial life).

If we adopt this perspective on life then we may draw several philosophical conclusions even before we get into the complexities of the thesis. The first general philosophical conclusion is that vitalists, reductive-mechanists, and emergentists fail to understand the machinery of life. The vitalist refrains from understanding the machinery of life by attributing it to a mysterious life force. The reductive-mechanist fails to understand that the machinery of life involves the orchestration of various components that exist on several distinct levels of analysis and that this orchestration cannot be reduced to simple functional rules. It must be mediated through signs. The emergentist fails to understand that life does not pop-up in a bottom-up fashion from simple interactions between microelements. Life is organized *in between* levels of analysis by laws of organization different from micro-level interactions. More specifically, the idea is that meaning is always created on the boundary of size scales (*e.g.* word–sentence–text, etc.) and an understanding of the machinery of life may benefit from studying this logic of in between. Therefore, the machinery of life should be studied through special attention to the mesoscopic level, or what has been described by Laughlin *et al.* (2000) as the "middle way", that is, a search for laws operating at levels and scales of organization (mesoscopic realms) intermediate between the microscopic state of fundamental particles and the macroscopic state of higher levels of organization. As argued by Strohman (2000):

> In biology, molecular genetic reductionism has mostly distracted us from study of mesoscopic realms between genotype and phenotype where complex organizational states exist and where as we now realize, there also exist networks of regulatory proteins capable of reorganizing patterns of gene expression, and much other "emergent" cellular behavior, in a *context-dependent* manner. (p. 575; emphasis mine)

I would like to argue that these rules are semiotic rules. Evolution should concern the appearance of in between dynamics as it is materialized in sign-mediated interaction. The last conclusion is that in order to understand life as a whole without falling prey to the vitalistic conception, we should develop new biosemiotic methodologies to understand how micro-elements are *orchestrated* through various communication channels to yield a functioning whole. This is not a simple task. The establishment of Darwinism as the central dogma of theoretical biology presents the skeptical scholar with a problem. Any potential contribution to the "grand narrative" of theoretical biology is limited *a priori* by the dogma to technical peculiarities. Indeed, rivers of ink have been spilled in an attempt to present, clarify, refine, and defend the different aspects of Darwinism and neo-Darwinism. The scholar who is fed up with technical particularities may suggest an alternative to this grand narrative. However, in such a case, where an alternative is suggested, it is immediately suspected of having a hidden creationist or vitalist agenda and ultimately dismissed as non-scientific. This situation should concern anyone who is open-minded enough to realize the difficulties associated with the theory of evolution.

Modern evolutionary theory suggests that inherited random genetic mutations and environmental selective pressure are the source of the observed variety of living creatures. According to this perspective evolution can be thought of as changes in the structure of DNA. Notice that according to this perspective the organism is in a tragic position between mutations that take place at the DNA level and brute forces that take place in its environment. Approaching life as meaning making we are opening a new spectrum of observations of life as an active and creative process of interactions through symbolic mediation. Life is about meaning. Does it mean that life is endowed with a "hidden spirit"? Am I presenting neo-vitalism? The answer is negative but the next chapter aims to present a point for thought about the physical grounding of meaning or about the "spirit" in the matter.

Chapter 12

A Point for Thought: Meaning—Bridging the Gap between Physics and Semantics

Summary

Attempts to apply information in its syntactic sense to biology encounter the sharp criticism of irrelevance. Nevertheless, when biologists reflect on their subject matter they inevitably invite bridging concepts, such as information, the relevance of which is not always clear. In this chapter, it is suggested that meaning, rather than information, is the appropriate concept for biologists, not only because as a new organizing concept it introduces new research questions but also because, in contrast to information, it can bridge the physical–biological–semantic gap.

1. Introduction

> There is no clear, technical notion of "information" in molecular biology, It is little more than a metaphor that masquerades as a theoretical concept and ... leads to a misleading picture of possible explanations in molecular biology. (Sarkar, 1996, p. 187)

> Biology deals with concrete components and processes: cells, molecules, binding, apoptosis and so on. However, biologists insist on using semantic concepts to describe their subject matter, for example, to describe genes and what they do. This is not a problem if there is a suitable "bridging concept" that links physical and semantic descriptions existing at two different logical levels of analysis. According to some philosophers and biologists, the concept of information "can play exactly that role". (Godfrey-Smith, 2002, p. 3)

Godfrey-Smith argues that biologists insist on using semantic concepts to describe genes and what they do. The inevitable question is "Why?" Can't biologists be satisfied with just their own biological data? Well, some biologists do want to settle for their own biological data. A nice anecdote illustrates this naïve position. Howard H. Pattee (2001) tells us the following amusing story:

> In the 1970s, a prominent molecular geneticist asked me, "Why do we need theory in biology when we have all the facts?" (p. 8)

The answer to this naïve question is also the answer to the question why biologists insist on using semantic concepts to describe genes and what they do. The answer is clear. Facts do not exist in isolation from theories, communication, and contemplating minds. The only mind that might hold bare facts without a theory, and I doubt this too, is the mind of the *idiot savant*, who may hold in his mind an enormous amount of what the psychologists call *declarative knowledge*, facts, without any organizing framework. In all other cases, facts need an organizing framework that exists at a higher level of analysis, which mean that the components of this level are *about* the components of the lower level (*i.e.* the "facts"), hence exist at the semantic level. In other words, there are no facts without a theory and we can go further and argue that there is no theory without a meta-theory. The shift from "it" to "aboutness", the semantic shift, is inevitable. In sum, to understand facts that exist on one level of analysis we must shift to a higher level of analysis. This higher level allows us to speak *about* facts and therefore it necessarily implies a shift from the physical level to the semantic level. There are people who may not accept this conclusion. For example, arguing against the use of "information" in biology, Boniolo (2003) writes:

> To give a correct and comprehensive analysis of the process which leads to the synthesis of the amino acid sequences a purely bio-physical approach is sufficient. (p. 259)

However, even Boniolo realizes that in order to explain "well known facts about the structure of the genotype and the gene expression" (Boniolo, 2003, p. 259) he should shift to another level of analysis. Boniolo chose the probabilistic language in order to explain these biological "facts". As we all know, probability is not a biological or physical concept. Probability theory is an extremely powerful mathematical system that we may sometimes use to *represent* our ignorance of the world (as argued by the subjectivist approach to probability) or to represent the erratic "nature of nature" as argued by others. In both cases, probability is not a part of the bio-physical realm. It does not exist in time and space as physical objects do. It is a way of representing a system and its behavior. It is a map that may help us to say something interesting *about* the territory. "Aboutness" cannot be avoided and, therefore, neither can the semantic shift accompanying it.

At this point and after we have realized that a shift to the semantic level is inevitable, we can start to examine different bridging concepts and their usefulness as bridging concepts. In other words, we should examine the relevance of different maps. More specifically we can ask what work these maps do and whether this work has some benefits for science. This is exactly the place where the concept of information gets into the picture. *Information*, like all the signs we know, has several senses. One of the most influential senses in which information has been discussed in biology is the syntactic sense as defined by Shannon.

Mathematical information theory studies the amount of information in a physical system, which is roughly the amount of order in the system. Is this sense of information relevant to the genetic realm? In his classical sense of information Crick (quoted in Boniolo, 2003, p. 258) argued that by information he means "the specification of the amino acid sequence of the protein [from the nucleotide sequence of the DNA]". At least concerning the genetic realm and Crick's definition of information, it is agreed that: "the syntactical approach to information is not able to grasp Crick's definition" (Boniolo, 2003, p. 258). Indeed, a crucial difference between information in biological systems and information *à la* Shannon is that information, as defined by Shannon, does not depend on the *sequence* of the subunits while biological information is defined precisely by that sequence (Barbieri, 2004). As suggested by Barbieri (2004):

> Physical information, in other words, has noting to do with specificity, while biological information has everything to do with it. (p. 92)

Although biological sequences cannot be characterized by Shannon's definition of information, there are other statistical concepts and tools that might help us expose the order of words in sentences or the order of genes in a sequence. For example, Markov models are used in natural language processing (NLP) when one can think of underlying events probabilistically generating surface events (Manning and Schutze, 2003). In this context, the criticism of Shannon's concept of information should be discussed in terms of the relevance of this concept for the questions we would like to answer.

2. Information: Different Questions Lead to Different Answers

The fact that we have a certain concept at hand does not mean that we have to use it or that it is relevant for our use. Unfortunately, fetish is a common perversion among scientists, and the concept of information, as introduced by Shannon, is one of the biologist's favorite fetishes. This idol of the mind can be illustrated in the context of immunity. In his introduction to "Forum: How Complexity Helps to Shape Alloimmunity", Orosz (2001) questions the reductionist approach, suggesting that the whole of a problem is but a sum of its parts and that an understanding of the whole can be achieved by an analysis of sufficiently dissected parts. His conclusion, which is not new, is that some problems resist such an approach. However, his example has instructional value for our case:

> If the question is what constitutes a watch? Or, how does a watch work?; then, the dismantling of a watch would be quite informative. However, if the question is: What is time?; then, dismantling a watch will probably provide little information of value. (Orosz, 2001, p. 1)

Molecular biologists or information-based theoretical biologists are sometimes those who ask about the structure of a watch. There are others who would like to study the function of the watch or how it works, and there are of course those who would like to study how a watch represents time and their question will necessarily lead them to questions of foundations such as "What is time?" Different questions are responded to with different answers and the philosophers' stone of information theory would not go far beyond its basic role of representing the entropy of a given system. In this context, information theory is, in many senses, irrelevant for understanding the interactions that constitute living organisms. The organism exists in time and its interactions-in-context should be studied, rather than their underlying genetic grammar. In this context, we do not have to accept Shannon's definition of information and we can draw on Bateson's naturalistic conception of information as a difference that makes a difference.

3. Information: A "Naturalistic" Perspective

Jablonka (2002) suggests that creating "a general definition of 'information' requires finding a common denominator for the different types of things or processes that we intuitively recognize as 'information' or as 'information carriers'" (p. 579). Her strategy, although problematic in several senses, is reasonable and her semantic definition of information is as follows:

> A source—an entity or a process—can be said to have information when a receiver system reacts to this source in a special way. The reaction of the receiver to the source has to be such that the reaction can actually or potentially change the state of the receiver

> in a (usually) functional manner. Moreover, there must be a consistent relation between variations in the form of the source and the corresponding changes in the receiver. (Jablonka, 2002, p. 582)

This definition presents a naturalistic conception of information which is highly similar to the one introduced by Bateson. To review, for Bateson (2000), information can be broadly considered *that which is conveyed by a message and provokes a response*, and a message can be considered *a portion of the world that comes to the attention of a system*, whether human or non-human. This general perspective on information opens up a field of interpretation that is broad enough for non-technical, scholarly discussions concerning the meaning of information and meaning in living systems.

Bateson (2000) defines a *bit* of information as "a difference that makes a difference" (p. 315). Simply stated, information is a differentiated portion of reality (*i.e.* a message), a piece of the world that comes to our attention and results in some response (*i.e.* meaning). In this sense, information is *interactive*. It is something that exists *in between* the responding system and the differentiated environment—whether the external or the internal environment.

For Bateson there is no clear difference between information and meaning, he considers them synonymous (Bateson, 2000, p. 130). Although it is true that information and meaning are intimately related, as Bateson suggests, they cannot be reduced to each other. Indeed, the term the *dual code* (Hoffmeyer and Emmeche, 1991) has been used specifically to express this idea.

Bateson presents the idea that a differentiated unit (*e.g.* a word) has meaning only on a higher level of logical organization (*e.g.* the sentence), that is, in context (Bateson 2000, p. 408), and only as a result of interaction between the organism and the environment. In this sense, the internal structure of the message is of no use in itself in understanding the meaning of a message (p. 420). The pattern(s) into which the sign is woven and the interaction in which it is located is what turns a differentiated portion of the world into a response by the organism. This idea implies that turning a signal (*i.e.* a difference) into a meaningful event (*i.e.* a difference that makes a difference) involves an active extraction of information from the message.

Bateson's naturalistic concept of information/meaning portrays a dividing line between living and non-living systems. As we previously suggested, meaning is the effect produced via semiotic mediation. How meaning is produced is still not clear. What is there in the non-mechanistic interaction that produces significance? To address this question we should be familiar with interactive machines and with a famous demon.

4. Interactive Machines

The statement "An organism is computed from DNA" is the bread and butter of modern biology. In this context, the genetic system can be approached as a Turing machine (TM) that transforms a finite input string of the genetic alphabet (*i.e.* DNA) into an output string (*i.e.* RNA and Proteins) through a sequence of state-transition instructions. This idea, popular and powerful as it is for certain aims, is extremely limited when trying to explain the *interactive* behavior of the genetic system. We have discussed the limits of the syntactic approach and these limits should be recalled as background for further analysis.

The general argument that a TM is limited in explaining the interactive behavior of systems has been presented in different versions and in various disciplines such as computer science (Wegner, 1998) and theoretical biology (Cohen, 2000a; Hoffmeyer and Emmeche, 1991). In the context of computing, Wegner (1998) provides a detailed argument explaining the limitations of a TM to model interactive behavior. The main limitation is that a TM is provided with an initial input but cannot accept external input while computing. In addition, the behavior of a TM is determined in advance and is stable across all possible worlds (Cleland, 2004). Therefore, it cannot model an interaction that is, by definition, communicative and a context-sensitive event. For example, a TM assumes that state-transitions operate on a string of a given finite alphabet. The translation from DNA to proteins involves a transformation from one set of a finite alphabet (the four letters of the nucleotides) to a different finite set of a finite alphabet (the 20 letters of the amino acids) through a *conventional* coding system (Barbieri, 2004). However, a TM cannot model transformations from one alphabet to another because it involves ongoing and continuous interactions with the surrounding cellular environment and extra-cellular environment.

Wegner suggests extending TMs to Interaction Machines (IMs) that interact directly with the external environment throughout computation. In fact, IMs interact while they compute (or compute while they interact) and *therefore their interaction can be considered in itself as a form of computation.*

An IM involves processes that are parallel in time and distributive in space. These processes interact between themselves and with input from the environment. Therefore, an IM cannot be reduced to a TM or realized by computable functions simply because the interaction yields new forms of behavior not expressible by mappings of strings (*i.e.* by functions). As argued by Wegner (1998):

> Functions are too strong an abstraction that sacrifices the ability to model time and other real world properties in the interests of formal tractability. (p. 317)

A virus interacting with an external unpredictable environment epitomizes this argument by ongoing adjustments to new input. The flexibility of the virus cannot be reduced to algorithms.

A key concept for understanding the behavior of an IM is *coordination-under*-constraints. I argue that coordination-under-constraints is an important aspect of meaning making. For example, when we argue that the meaning of a given molecule being an antigen is determined in context (Cohen, 2000a), we actually argue that the coordinated behavior under constraints of the immune agents generates the decision whether the molecule is an antigen or not. Meaning making involves coordination-under-constraints.

Biological interaction/computation can be considered as a form of measurement. Let me explain this claim. When two biological entities interact they do not interact only according to purely physical forces. Their interaction has an informational value. When ligand–receptor binding takes place this binding constrains the potential conformations of the protein and moves the protein molecules into a certain *basin of attraction*. This process, which can be described in physical terms, is a process in which information is produced in the sense that the constraints imposed on a "problem space" are used for producing a response. It is a process in which constraints are imposed on the behavior of a system. If this information can somehow be used for producing information on a higher level of analysis, then we can say it has turned into meaning—a difference that makes a difference. To better understand this process, in which interaction/computation/measurement produces information, let us turn to the assistance of a demon—Maxwell's demon.

5. Maxwell's Demon

> "*Which demon has leapt further than the largest leap?*" (Oedipus Chorus)

In our minds, demons belong to the pre-scientific era in which supernatural beings, malevolent spirits, populated the human world. Maxwell's demon is the physicist's favorite demon, just as the *Escherichia coli* is the biologist's favorite pet. However, Maxwell's demon is not a vicious spirit. It is the hero of a thought experiment invented by the famous physicist James Clerk Maxwell in 1871. In the introduction to *Maxwell's Demon: Entropy, Information and Computing* (1990), the editors, Leff and Rex, describe Maxwell's demon as an idea that challenged some of the best scientific minds. Who is the demon that challenged "some of the best scientific minds?" To address this question we should understand that the demon was presented in the context of the second law of thermodynamics. The second law suggests that the total entropy of any isolated thermodynamic system tends to increase over time, approaching a maximum value. The second law is a statistical law and thus applicable only to macroscopic systems. When one part of an isolated system interacts with another part, energy tends to distribute equally among the accessible energy states of the system. As a result, the system tends to approach thermal equilibrium, at which point entropy is at a maximum and the free energy which can do some work is zero. For example, let us imagine a box that includes two chambers separated by a barrier. In one chamber we have molecules of gas, and the other one is empty. If we remove the barrier the molecules will spontaneously move in the 3-D space of the box and their erratic movement will result in a random organization of the molecules in the space. From a state in which the molecules were concentrated in one part of the box, we get a state in which the molecules are disordered. The entropy of the system has been spontaneously increased. In a closed system which is not in a state of equilibrium, as time unfolds entropy increases. The ticket from order to disorder is free, but for the other way around, one has to pay by doing some work and investing some energy.

Maxwell's demon has different interpretations but in general it is conceived as a thought experiment that questions the second law or illustrates its statistical nature. The demon is an extremely small creature that sits in our box and sorts molecules by their velocity with little work and dissipation. In Fig.12.1, you can see an illustration of the demon at work. When the demon notices a fast molecule he opens the barrier and allows it to get in. When he observes a slow molecule it is left out. After a while, and in contrast with the second law, order will be almost spontaneously generated from disorder.

Fig. 12.1 A schematic representation of Maxwell's demon.

To decide which molecule is slow and which molecule is fast, our demon has to be intelligent enough to conduct a *measurement*, in our terms, to turn a difference (in velocity) into a difference that makes a difference, into a response, into meaning. Unfortunately, it seems that our creature cannot produce order from disorder without paying the price in energy and therefore the second law is not violated.

We can convert the system's entropy to known information by measurement, and known information into entropy by forgetting or erasing it (Frank, 2002). Saying that we can convert entropy to known information through measurement seems like a contradiction of the second law of thermodynamics. However, if we can measure a system then it is not completely closed from an informational standpoint. This is an important point. *Measurement always assumes a shift from a given system to a meta-system.* Turning a difference into information, and information into meaning involves a movement between different scales of analysis, and a *semantic shift*. Therefore, measurement is possible only in open systems and only by imposing constraints on the *firstness* (potential) of the lower levels. Now, we can understand why meaning making is the defining characteristic of living systems, a point that will become clearer as this chapter unfolds.

A second issue is the precedence of meaning over information. A molecule can be considered as fast or slow only if the demon can make comparisons. This is the paradox of information. Without a preceding scheme of meaning, information is "informationless". This statement does not necessarily entail a regression of demons sitting in the minds of demons and the paradox can be settled if we understand that the topology in which our demon operates is the topology of the *Klein bottle*. In the final section of this chapter I will try to explain how to save our demon from another demon—the demon of vicious regression. For the time being the issue of measurement and information should be better elaborated.

I brought up Maxwell's demon only to illustrate a certain point: that *measurement creates order* and that measurement collapses an indeterminate signal into value. This point will be elaborated upon in a chapter concerning the polysemy of the sign. Information, in the sense of differentiation, must assume an intelligent device that invests energy to turn disorder into a differentiated realm. The idea that measurement generates information is highly relevant for biologists for one major reason: *Meaning is created when a given system's degrees of freedom are constrained in a regular way for a given observer (i.e.* another system that use the regularity as an input for its maintenance and functioning). Our demon does not constrain the entropy of the system for the sake of amusement. It constrains the system's degrees of freedom for the sake of a very concrete function. Our demon is itself subject to measurement by another demon that produces a difference that makes a difference, and these demons in themselves are immersed in a recursive-hierarchical network of demons that constitute the living organism. We will elaborate upon this point further by discussing metabolic networks and the strange question: What happens when Maxwell's demon is constructed as a Klein bottle? First, let us better clarify the meaning of measurement and the relation between measurement, information, and meaning.

6. Measurement as an Invention

Measurement Science has been defined as

> the systematic study and organization of the methods by which information is gathered from the physical world. (Cropley, 1998, p. 223)

However, it is meaning that stands at the heart of measurement science and not information *per se*. As argued by Cropley (1998):

> The central qualitative issue in a study of measurement information is that of *meaning*. (p. 229; emphasis mine)

Let us try to understand this argument. First, the idea that measurement generates information dismisses the naïve conception that information has some kind of Platonic status, which is ontologically prior and epistemologically indifferent to any kind of observer or measuring process. This naïve conception has interesting historical roots. Historically, people start measuring things by dealing with a small number of well-defined objects. Measurement was intimately related to the procedure of counting and with natural numbers. The shepherd counted the number of sheep in his herd, the leader of a tribe counted the number of his women, and so on. Numbers were used to represent countable objects and when they did not represent sensual, countable objects they were rejected or accepted with suspicion. This conception of numbers impeded the conception of measurement because it assumes the object or the objects of measurement should be identified prior to the measurement and correspond to observed entities. However, this conception has been deeply transformed. Today, we realize that *measurement precedes and determines the object and not vice versa* (Algom, 1986). For example, we determine the weight of a certain object and its height above the ground, multiply the two entities and name the result "potential energy". Potential energy is the product of our measurement; it is not something that walks around in our yard. As argued by Algom (1986) *measurement is an invention and not a discovery*.

The justification of a measurement is not the revelation of a hidden object but the theoretical or practical benefits of the measurement. To recall Bakhtin, "there is no alibi in existence" and the justification of measurement is practical. The reason why the idea of measurement is relevant for biologists concerns the relation between meaning and information. Measurement produces information from entropy but information is not enough. The information encapsulated in DNA is worthless unless it is translated to proteins. Can it be that meaning is created by a second-order measurement process?

Let us summarize the argument so far. Living systems are not TMs but IMs, or following Bakhtin, we may call them *dialogical machines*. Their defining characteristic is interaction mediated by signs—communicated functional generalities. This form of interaction involves a measurement—a process in which indeterminate signals turn into a differentiated realm that exists at a higher level of analysis. This activity encapsulates a quandary of regression. A difference, in velocity for example, can make a difference only for a demon, a measurement device that already has in its mind the different concepts of fast and slow. How can we save Maxwell's demon from the demon of infinite regression? How can we avoid a situation in which inside Maxwell's demon's head sits a little homunculus, a demon in itself, that has in his mind another demon-homunculus, and so on? The next section aims to provide a solution and to argue that biological systems compute themselves in a unique way that avoids the confrontation with the demon of infinite regression.

7. Maxwell's Demon in a Klein Bottle

An amusing fable (Gardner, 1996) tells us that a well-known scientist once gave a public lecture on astronomy. The scientist presented to the audience the structure of the universe, including the way the earth spins around the sun, and the way the sun itself spins around the center of our galaxy. At the end of the lecture, an old lady sitting in the audience got on her feet and said: "All of what you have told us is pure nonsense. The world is a flat board that rests on the back of a giant tortoise". The scientist smiled patronizingly and asked the old lady, "And what does the tortoise stand on?" "You're very clever, young man, very clever", replied the old lady, "but its turtles all the way down!" In her answer the old lady epitomized the demon of vicious regression. This demon is evident in a measurement process in which the observer is also the outcome of the measurement process. This vicious regression is solved if the measurement is taken place along the lines of the Klein bottle. Let us explain what the Möbius strip/band and Klein bottle are.

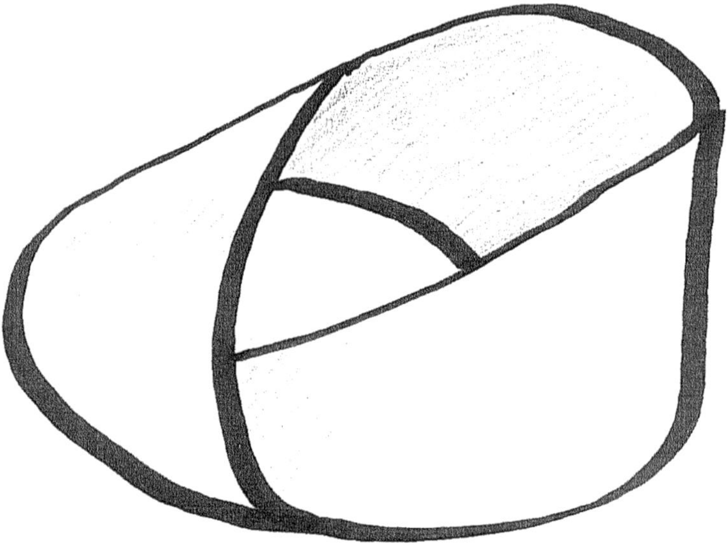

Fig. 12.2 The Möbius strip.

The Möbius strip/band is a one-sided surface in the sense that a bug can traverse the entire surface without crossing an edge (*i.e.* a point of discontinuity). Figure 12.2 is the sketch of the Möbius strip/band.

The Möbius band is interesting since it is non-orientable. In geometry and topology, a surface is called non-orientable, if a figure such as the letter R can be moved about on the surface so that it becomes mirror-reversed. Otherwise, the surface is said to be orientable. A non-orientable surface may allow us to restore the symmetry of an object sliding on it. It is an example of a topology that may allow symmetry restoration.

Another example of a non-orientable surface is the Klein bottle. The *Klein bottle* is a higher-dimensional topological version of the Möbius strip (Fig. 12.3). It is, roughly speaking, the product of two Möbius strips glued together along each of their lone edges (a proper Klein bottle can only exist in four dimensions; it can be only imperfectly represented in three).

What is important to notice about the Klein bottle is that it is a topological structure that *passes through itself* so that outside and inside meet. It is a re-entering form. A bug traveling along the surface of the bottle does not cross a region of discontinuity and therefore cannot discriminate between inside and outside. By penetrating the bottle from higher dimensionality a bug can move from the inside to the outside without crossing a point of discontinuity.

Now, let us assume that an organism is built along the lines of the Klein bottle. This organism is constituted of a network of demons, each simultaneously measuring and being measured. In this case we have a system

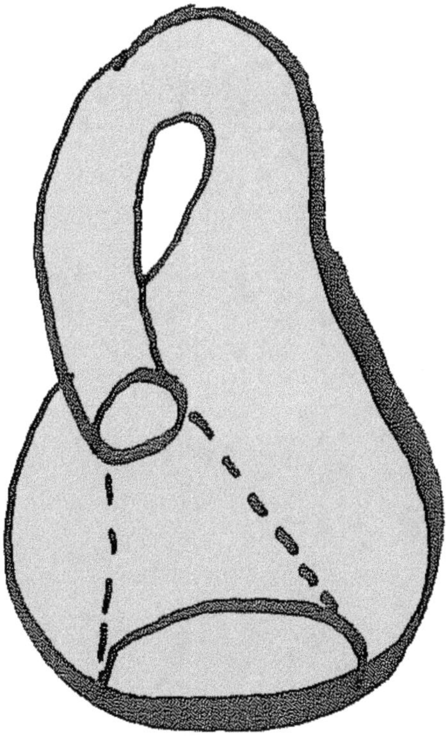

Fig. 12.3 The Klein bottle.

that observes itself without falling prey to the old lady's demon of infinite regression. This abstract idea will be later illustrated with regard to the idea of a recursive-hierarchy and the way it can explain the mystery of hidden life—cryptobiosis. In the meanwhile, we can imagine the scientist listening emphatically, rather than ironically, to the old lady and saying: "You are definitely right my lady. Indeed there are turtles all the way down. But the turtles exist all the way down in a Klein bottle". I can imagine the old lady's shock: turtles in a bottle?

Chapter 13

The Rest is Silence

In the previous chapters, we realized that neither the organization of the text's particles nor the correspondence between a sign and a signified are sufficient for determining the meaning of an utterance. Meaning is *the effect of interaction between at least two parties, and is produced via semiotic mediation.* An understanding of this effect must take into account environmental factors in which the communicated signals are embedded (context) and the inference process through which the effect is produced. This is exactly where *pragmatics* gets into the picture.

Pragmatics studies "the use of language in human communication as determined by the conditions of society" (Mey, 2001, p. 6). This definition emphasizes the *use* of language rather than its *structure*, *communication* rather than *organization*, and language as a *context-dependent* and context-oriented activity rather than a *context-free* activity. Context is the label we give to the totality of these environmental factors, and understanding context is crucial for understanding meaning, as I have repeatedly emphasized. In this chapter, I would like to deepen our understanding of meaning making by introducing pragmatics—the field of linguistics that deals with the way meaning is generated in context. Let me open this chapter by clarifying the meaning of context.

In his paper, "Context in Context", the social historian Peter Burke (2002) presents a condensed social history of the term. *Contexere* is classical Latin for "to weave" and the noun *contextus* is used in the sense of connection. Later, in the 4th century A.D. the noun *contextio* came into use. This noun describes the text surrounding a given passage that one wishes to interpret (Burke, 2002). The term was used later in the 16th and 17th centuries in the context of the interpretation and translation of texts. For example, the practice of translation raised the need to attend not only to the individual words, but also to the relation between them. Thus context in its original material sense of weaving was expanded metaphorically to include various senses such as the text surrounding a word, the coherence of the entire text, and the intention of the writer of the original text. Burke points out that the use of *context* appears in a context of opposition against various reductive structuralists who were trying to argue for the internal, structural, or universal justification of knowledge claims: for example, scriptural fundamentalists who believe that eternal wisdom is encapsulated in the holy texts, or genetic reductionists who believe that the organism may be reduced to a genetic sequence.

Context is "the quintessential pragmatic concept" (Mey, 2001, p. 14) and may be extended beyond the realm of human language by using the above definition of context. In this context, it is important to reject the naïve conception of context as a passive background for the interpretation of signs. Context is not the stage on which the drama of communication unfolds. As suggested by Mey (2001):

> Context is action. Context is about understanding what things are for; it is also what gives our utterance their true pragmatic meaning and allows them to be counted as true pragmatic acts ... (p. 41)

This statement emphasizes the fact that a context turns into a context not by being passively given but through the active involvement of the agent. If an agent actively uses context, can we determine its boundaries? This general question concerning the boundaries of context was faced years ago by Gregory Bateson. The example he used and his conclusion are illuminating:

> Suppose I am a blind man, and I use a stick. I go tap, tap, tap. Where do *I* start? Is my mental system bounded at the handle of the stick? Is it bounded by my skin? Does it start halfway up the stick? Does it start at the tip of the stick? But these are nonsense questions ... The way to delineate the system is to draw the limiting line in such a way that you do not cut any of these pathways in ways which leave things inexplicable. If what you are trying to explain is a given piece of behavior, such as the locomotion of the blind man, then, for this purpose, you will need the street, the stick, the man; the street, the stick and so on, round and round. (Bateson, 1973, p. 434)

Bateson's suggestion is that the boundary of a system is fuzzy and cannot be determined *a priori* and outside the situation, the same as the boundaries of an utterance. The boundary of the blind man may include artifacts such as the stick. The stick is a part of a context that should be taken into account in order to explain the locomotion of the blind man. As suggested by Burke (2002, p. 172), "What counts as a context depends on what one wishes to explain".

1. Wound Healing, Plasticity, and Context

The ability of the organism to attune to context is sometimes described in biology in terms of *plasticity*. Plasticity illustrates the embedded nature of context because plasticity in itself is subject to contextual factors. For example, our developmental trajectory is such that we are more plastic at the beginning and at a certain phase lose our basic plasticity for a relatively more stable form. In other words, plasticity is contextualized. The contextual nature of our ability to attune to context (*i.e.* plasticity, or the embedded nature of context) is illustrated by a comparison of wound healing among embryos and mature organisms. Let me explain. Our body is protected by the skin, which serves as our boundary. When this boundary is breached there is a danger of losing liquids that are crucial to cell functioning and a danger of invading microbes. Keeping the integrity of the tissues is therefore a major task facing the organism.

On the single-cell level it is known that a vertebrates' cell maintains its integrity against mechanical pressure by using the cytoskeletal network, a network of proteins that provide the cell with scaffolds. Another protein that is important in maintaining cell integrity is Vimentin. It has been shown that connective tissues (fibroblasts) deficient in Vimentin were 40% less stiff than wild-type cells (Woolley and Martin, 2000). The interesting thing, however, is not the role of certain proteins in constituting the cell's resistance against mechanical damage, but the process of *wound healing*.

On the single-cell level the entry of calcium into the cell signals a problem. This signal triggers the fusion of internal vesicles, membrane shells, with each other. This membrane bilayer blocks the breach in a way similar to that used by sailors who close breaches in ships using sheets. Multi-cellular wound healing is more complex. Re-epithelialization is accomplished in different ways in the adult and the embryo. In adults, wound closure is accomplished through the active movement of connective tissue and epidermis. The connective tissue contracts to pull the wound edges together and the epidermis then moves to cover the exposed connective tissue. More specifically, cells within the front rows crawl or extend finger-like thin structures (*i.e.* lamellipodia) that extend and retract, and in so doing cover the wound.

In embryos *wounds heal rapidly and perfectly without leaving a scar* (Redd *et al.*, 2004). In contrast, wound healing in the adult is imperfect and results in scars. What is the mechanism that allows the embryo to repair damaged tissue with such perfection? Embryos also use the adult wound healing mechanism, BUT the cellular mechanisms for movement of connective tissue and re-epithelialization are quite different (Redd *et al.*, 2004). In the embryo lamellipodial crawling does not exist. It was found that a thick cable of actin at the leading edge of the marginal cells encompassing the wound is responsible for closing the wound. Contraction of this cable provides the force that draws wound edges together in a purse string-like manner (Nodder and Martin, 1997).

Inflammation is another important difference between wound healing in the embryo and the adult. At the site of the adult's wounds inflammation is evident while in the embryo wound healing process inflammation is minimal to non-existent. In the adult's wound the inflammatory cascade is activated by release of cytokines and growth factors from degranulating platelets. A platelet is a particle found in the blood stream that binds to fibrinogen at the site of the wound and begins the blood clotting process. However, in the early embryo there are no platelets, and therefore the basic trigger for the inflammatory response simply does not exist. Growth factors seem to be a crucial factor in embryo wound healing and controlling growth factors in adults was shown to be associated with reduced scarring at the wound site (Redd *et al.*, 2004).

Why is wound healing among the embryo more rapid and efficient? The answer I would like to provide is rather simple. The embryo is in a relatively higher state of transition. Plasticity is implied from this higher rate of transition in which the organism's self is ill defined and should resonate with a changing context. In other words, *plasticity, the ability to attune to context, is in itself context-dependent.*

The fact that context is embedded and hierarchical, that a context has fuzzy boundaries, and that it is dynamic does not mean that a contextual analysis of a communicative event, whether among human beings or other biological systems, is impossible or meaningless. The next sections aim to take us from the realm of linguistic pragmatics to the realm of "biological pragmatics". By inquiring into the phenomenon of silencing, I will show the similarities between silencing in human communication and silencing in biological processes, and the benefits of considering this phenomenon from a pragmatic perspective, that is, the benefits of considering silencing from the perspective of meaning making.

2. Epigenesis

The theory of evolution, specifically from the molecular perspective, is a powerful theory for explaining the *structural* variety of living creatures. However, it has several major difficulties. The first difficulty is that the theory of evolution does not provide us with any criterion for differentiating between living and non-living matter, and does not give us a clue about the origin of life on earth although it has some serious suggestions about how to trace back the phylogenetic tree of living beings. The theory of evolution does not explain to us in what sense organisms are different from matter and it is usually associated with the idea that life on earth originated from random interactions of physical particles.

Another problem with the theory of evolution is with the one-directional flow of influence from genotype to phenotype. This one-directional flow encounters opposition from recent findings concerning epigenesis. These findings suggest that the environment plays the role of a landscape that channels the potentialities of the genome. This idea does not necessarily dismiss the theory of evolution but just adds another layer of complexity to the process. However, taking it to its extreme, epigenesis may be used as a serious challenge to neo-Darwinism (van Speybroeck *et al.*, 2002). If the environment interacts with the *potentialities* of the genome, then a shift of focus should be made from (1) considering the genes as the mechanistic determinants of the phenotype *à la* Dawkins to the genes as the tokens that the organism uses for self-creation, maintenance, and reproduction, (2) the focus on random mutations to built-in potentialities, and (3) a focus on the environment as a selective device to the environment as a landscape for ecological interactions.

A contextual approach transforms the role of the environment from a *strainer* to a *landscape*. In this sense, the environment is a nested set of contexts in which development, life, and growth take place. According to this conceptualization the organism is in constant dialogue with a variety of environmental factors. This dialogue produces meaning, and a meaning-making perspective should illustrate the way in which this meaning is produced in context. The following text explains epigenesis and shows how is it associated with silencing: Epigenesis is a term used to describe the

> interactions through which the inherited potentials of the genome become actualized into an adult organism. (Gilbert, 2000, p. 202; emphasis mine)

Gilbert's definition hides a conundrum. What are these inherited potentialities of the genome? Are these potentialities just the syntax of the genome triggered by environmental cues? This is an important question because it concerns the issue of biological plasticity in general and developmental plasticity in particular. Contextual cues fold (imply—*plicare*) potential meaning into what Volosinov describes as the "theme".

Developmental plasticity (or phenotypic plasticity) is the idea that the genome "enables the organism to produce a range of phenotypes" (Gilbert, 2001, p. 3). Polyphenism is a specific instance of this plasticity and it refers (within a single population) to the occurrence of a discontinuous phenotype elicited by the environment from a single genotype. For example, temperature is an environmental factor (*i.e.* a contextual cue) that influences the phenotype of the European Map butterfly. In summer the phenotype is black with a white band and in spring it is bright orange with black spots (Gilbert, 2001). Temperature is an important contextual factor for the evolving Map butterfly. Another example of polyphenism is the influence of temperature on the sex of the turtle embryo: At one temperature it becomes a male and at another temperature it becomes a female. In this example, too, temperature is an important contextual factor. The genome is the same but it responds flexibly, or more accurately, it is interpreted *flexibly by the organism, through contextual hints*, to produce different phenotypes.

One should not consider these cases as zoological obscurities, but rather, remember that the plasticity of the genome and its reactivity to signals from the environment is what underlies the rich repertoire of the adaptive immune system.

As was argued by Gilbert (2005, p. 65): "environmental context plays significant roles in the development of most all species". It is also argued by Gilbert (2005) that environment can influence the development of the organism by at least three major routes: (1) the neuroendoctrine system, (2) as an embryonic inducer, and (3) as a transcriptional modulator that effects genes expression through methylation. In all of these cases the environment plays an active role, which is different from what was traditionally attributed to it by evolutionary biologists. I will briefly present one way in which the environment influences the phenotype through methylation.

Methylation is the addition of a methyl group to any substrate. A methyl group is a hydrophobic ("water hater" or oily) group, which is composed of carbon and hydrogen atoms (CH_3). In the case of epigenesis the term refers to the addition of a methyl group to cytosine, one of the basic letters that comprises DNA. Methylation occurs at the relatively rare sites in which cytosine is located near guanine and separated through a phosphate (*i.e.* CpG). Although the neighborhood of cytosine and guanine is rare, it is more frequent in areas known as "CpG Islands".

Cytosine's chemical formula is $C_4H_5N_3O$ and methylation occurs when a methyl group is attached to carbon-5 and alters the properties of the base. The methylation of these sites is crucial for gene activity/expression. It is argued that DNA methylation has several functions. One of the major functions is to silence repetitive sequences (parasitic DNA) that have entered the genome via viral infection through evolution (Allegrucci *et al.*, 2005). In other words, DNA methylation is an epigenetic and meta-genetic function that aims to assure appropriate gene expression. This is why epigenesis may be also defined in the genetic context as alterations in DNA function without alterations in DNA sequence (Jones and Takai, 2001). The sequence has not been changed but the function is changed. In this sense, pragmatic meaning reigns over genetic grammar.

The fact that the function of DNA can be altered without causing a change in its sequence illustrates the plasticity of the genome but also points to the important role of meta-regulatory processes that cannot be simply reduced to DNA sequences. Turning again to the linguistic metaphor, we may argue that methylation is a pragmatic function of "genetic language". It does not change the sequence but regulates its function. It is a meta-linguistic activity. Here we encounter a problem. The role of DNA methylation is to silence junk DNA. However, as reviewed in a previous chapter, it was recently argued that what was described as junk DNA, or at least the portion of it known as ncRNA, is responsible for the methylation process! In other words, the basic capacity to control gene expression is encapsulated in the same places the methylation process is supposed to silence. *Recursion defies linearity*! To illustrate the recursive and pragmatic nature of biological processes I will present silence in natural language, discuss silencing in genetics, and conclude with the recursive way in which biological systems maintain their existence. Meaning, interaction, semiotic mediation, context, and recursive-hierarchy are central keywords crucial for understanding living systems as meaning-making systems.

3. The Rest is Silence

In his essay "The Rest is Silence", Huxley (1931) writes:

> From pure sensation to the intuition of beauty, from pleasure and pain to love and the mystical ecstasy and death—all the things that are fundamental, all the things that, to the human spirit, are most profoundly significant, can only be experienced, not expressed. The rest is always and everywhere silence. (p. 19)

Huxley is probably correct in the sense that silence marks the boundary of our understanding. Nevertheless both language and silence are crucial for understanding, as suggested by Ortega y Gasset:

> The stupendous reality that is language cannot be understood unless we begin by observing that speech consists above all in silences. (Ortega, quoted in Becker, 2000, p. 285)

Anton Becker (2000) points at Ortega's insightful observation into language. Ortega noticed that in language we have a delicate balance between manifestation and silence and that "each people leaves some things unsaid *in order* to be able to say others" (Becker, 2000, p. 6). If silence is an important part of language, or more accurately of the language activity, what Becker, following Maturana and Varela, calls "languaging", then when translating between languages we are facing a problem. Not only do we have to translate what is said but also what is unsaid. This statement holds true both for language and for biology, where silencing of certain genes is a crucial activity of the working system. How can we translate the "unsaid"? How can we translate a silence? How can we translate that which has no grammar?

Becker's answer is that in translation there are things necessarily left behind. In this sense, a complete translation is impossible. For example, the simple English expression "I am" is untranslatable to Burmese, which has no neutral first-person pronoun, no tense, and no copula. How can someone argue that there is a universal grammar of the mind when a major constituent such as the neutral first-person pronoun does not exist in Burmese?

The difficulty of silencing caused Ortega to suggest that a "theory of saying, of languages", would also have to be a theory of "the particular silences observed by different people" (Ortega, quoted in Becker, 2000, p. 285).

In human communication silence has different senses (Jaworski, 1997). It can be a tool for manipulation, an expression of taboo or repression, or an expression of artistic ideas. Silence can be linguistic, kinetic (stillness), but also biological. There is no point in restricting silence to the linguistic realm *per se* and it can be extended to the biological realm. We may consider this extension as an intellectual wisecrack by academics. However, when we turn from the linguistic realm to the genetic realm we immediately realize that silencing is a practical biological issue of life and death. Let me explain why by discussing gene silencing in plants.

Plants are exposed to the harmful invasion of viruses. Viruses are the ultimate expression of parasitism. They can reproduce only by using the machinery of cellular organisms. Unfortunately this parasitic activity might infect the host organisms. Here different defense mechanisms get into the picture. Viruses have RNA rather than DNA as their genetic material and through their life cycle they make double-stranded RNA (dsRNA). In this context, a possible defense mechanism of the organism may identify this dsRNA and take care of it. This is exactly what happens during RNA interference.

Silencing can occur at the post-transcriptional phase and it is called "post-transcriptional gene silencing" (PTGS). PTGS is evident in different kingdoms (fungi, plants, animals) and species (*e.g.* zebrafish, mice). In plants, where PTGS was first identified, it was found that it is the dsRNA, which is the factor that triggers PTGS. By experimenting with one of the biologist's favorite pet, the nematode *Caenorhabditis elegans*, it was found that dsRNA could lead to gene silencing.

Antisense RNA is a complementary RNA sequence that binds to a messenger RNA molecule and thus blocks its transcription. As we know binding can be lethal. The antisense RNA protein can mark viral genetic instructions and thus interfere with the reproductions of viruses. Antisense RNA is also a tool for assessing the function of a gene. If we shut down the expression of a certain gene we can observe the phenotypic consequences of this intervention and thus learn about the function of the specific gene. When two researchers (Guo and Kemphues, 1995) used antisense RNA to shut down the expression of a certain gene, they found something surprising. They found not only that the injection of the antisense RNA blocked the expression of the gene but that the injection of the sense-strand control did the same (Guo and Kemphues, 1995). That is, even when we introduce a naturally occurring mRNA molecule we silence the gene. Moreover, it was later found that the injection of dsRNA—a mixture of sense and antisense strands—created a stronger silencing effect. This result was puzzling. How is it possible that sense and antisense RNA create the same silencing effect? The logic of RNA interference will be explained below.

PTGS introduced by dsRNA is described as RNA interference (RNAi) and it has turned out to be a powerful tool for silencing genes, as well as a hot topic in biological research with great promise for the development of new drugs. In a review paper written by Hammond *et al.* (2001), the authors write that RNAi "shows several features that border on the mystical" (p. 10). Whenever biologists realize that their research borders on the mystical there is a place for the intellectual nomad to get into the game, but let us first understand the mechanism.

How does the RNAi work? I will use the paper and the illustration of Hammond *et al.* (2001) to explain this process (Fig. 13.1).

In the first step, input dsRNA is cut by an enzyme called dicer into small segments known as *small interfering RNAs* (siRNAs) or *guide RNAs.* We could have expected that this process would be the end of the dsRNA however in the effector step the siRNs bind to a nuclease complex, an enzyme that breaks the phosphate group that bonds nucleotide subunits, and form the *RNA-induced silencing complex* (RISC). This complex machinery uses the short pieces of the RNA produced by the dicer as

> a template to seek out and destroy single stranded RNA with the same sequence, such as mRNA copies used by the virus to direct synthesis of viral protein. (Downward, 2004, p. 1246)

This complex targets the homologous transcript by base pairing interactions and cleaves the mRNA nucleotides. The mRNA is destroyed and silence reigns.

One should realize, however, that things are much more complicated and obscure in higher-order organisms (mammals) where anti-viral activity relies

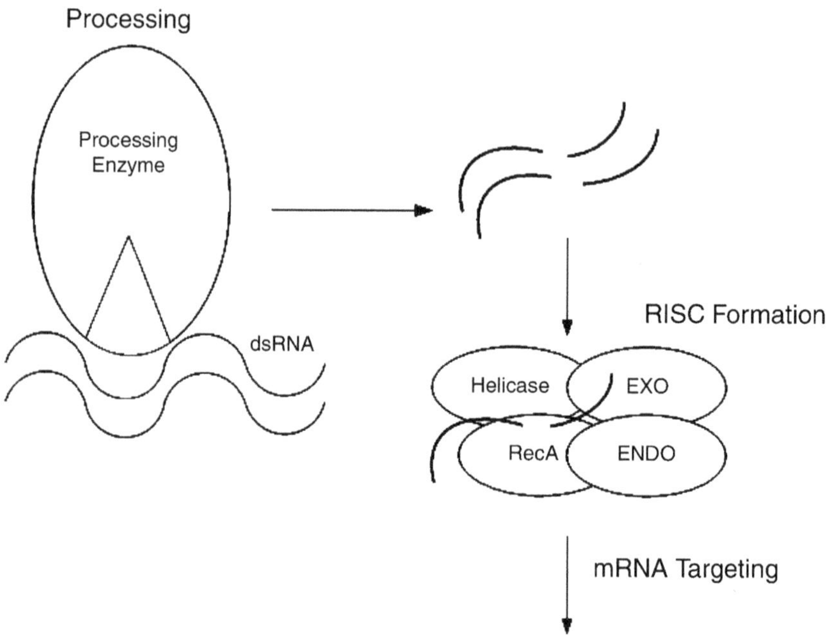

Fig. 13.1 RNA interference.

on the production of interferons resulting in the inhibition of gene expression and rapid cell death that prevents the replication of the virus in the organism (Downward, 2004). However, it seems that PTGS is an ancient defense mechanism against transposons (small mobile DNA sequences that can replicate and insert copies at random sites within the chromosomes) and viruses whose genetic material is RNA (Cogoni and Macino, 2000), and that RNAi exists across kingdoms to include mammals. In fact it was found that there are two different kinds of siRNAs, one involved in virus defense and the other that deals with transposons (Baulcombe, 2005).

Baulcombe (2005) describes siRNA as the "dark matter" of genetics because he suggests that we have to anticipate that "there are thousands of different siRNA molecules in a plant" (p. 26) and that siRNA plays other biological functions in addition to silencing.

An interesting theoretical question I would like to address in this chapter is why silencing? Why not simply destroying the naughty transposons, for example, and prevent them from propagating into future generations? In psychoanalytical terms we may ask why should we repress/silence the unconscious drives of the id rather than just get rid of them. Here the pragmatic-linguistic approach to biology may be valuable since it may provide us with a theoretical perspective to address this question. First, we should realize that the complexity of organisms entails silencing:

> In unicellular organisms, most of the genes in the genome are in a perceptual state of activity, with only a small number being specifically recognized as targets for repression. By contrast, repression is a dominant theme in the regulation of gene expression in animal cells, with more than 50% of the genome being silenced in any particular cell type. (Lande-Diner and Cedar, 2005, p. 653)

Silencing is clearly an epigenetic activity since it involves alterations in DNA function without alterations in DNA sequence. As such, silencing is also a contextual activity. Some genes are silenced during development in order to restrict the inherited potential of the cell to take different forms. "Cell plasticity is lost as development proceeds into adult life" (Lande-Diner and Cedar, 2005, p. 53). The loss of cellular potential "correlates with irreversible gene silencing" (Lande-Diner and Cedar, 2005, p. 53) as development proceeds. As reiterated by Meehan (2003), the loss of this highly orchestrated activity of gene silencing has lethal consequences in development.

Here, we come to the point where we can clearly understand why silencing is preferred to extinction. It is not that we do not need the silenced genes. *In certain contexts, we do not really need them*. However, in other contexts, we need them depending on the timing of their expression. In other words, *the complexity of multi-cellular organisms assumes context sensitivity* and silencing is just a particular case. The orchestra must play, but to turn the music into a symphony timing and silencing are necessary to avoid cacophony.

4. How Do Biological Systems Know Themselves?

An interesting issue concerns the reflexivity that characterizes gene silencing. To silence certain genes the genetic system must recognize self-characteristics that are not a part of the genetic self. This recognition activity implies self-knowledge and reflectivity, which are recurrent themes in this manuscript. However, reflectivity cannot be explained through a reductionist approach. Reflectivity and self-knowledge always involve a shift in perspective to a higher level of organization. This fact is realized by researchers in genetics, as is suggested by Meehan (2003):

> Throughout development, there are distinct patterns of gene expression set up in somatic cells, which are stably inherited through cell division. Consequently, there must be *a level of information apart from the primary DNA sequence that specifies the*

> *selective use of genetic information during development.* (p. 53; emphasis mine)

One should not mistake this statement to imply a non-scientific and spiritualistic perspective. *In vivo*, the self-knowledge of the genetic system is actualized in matter. There is nothing mystical or spiritual in this phenomenon in the oversimplistic sense of the terms *mystical* and *spiritual*, and one should acknowledge the reflexive behavior of the genetic system and realize that reflective behavior is not the sole property of the conscious human organism.

Self-knowledge is necessary but what happens when the self is not yet well defined, such as in the case of embryo development? The system does not know itself. How can it know itself without having a well-defined self? How can we sketch a map of a territory yet to be uncovered? Is there knowing without a knower? The answer to this mysterious question is probably positive. It seems that in contrast to our naïve homocentric conception, the logic of living systems is such that *knowing precedes the knower* and the map precedes the territory. The most well-known spokesman for this dynamic approach in biology is Brian Goodwin who considers "creative emergence as the central quality of the evolutionary process" (Goodwin, 1994, p. xiii). This perspective suggests that the nature of organisms might be grasped through a

> relational order *between* components that matters more than material composition in living processes, so that emergent qualities predominate over quantities. (p. xiv)

Goodwin (1994) suggests that ordered complexity emerges through a self-stabilizing cascade of symmetry-breaking bifurcations that have an intrinsically hierarchical property in which higher-order forms constrain the dynamics of the lower level that generates the shape (p. 100) The transition from a state of high symmetry (lower complexity) to a state of low symmetry (high complexity) is known as bifurcation (p. 89). Organisms, as evolving systems, have a "dynamic stability".

Goodwin is a structuralist, in the sense that he tries to understand biological forms by inquiring into the abstract structure/dynamics that constitutes a form. The thesis presented in this book clearly resonates with Goodwin's perspective. Unfortunately, in his well-known book Goodwin does not mention a brilliant structuralist—Jean Piaget—whose depth of thinking and its general relevance has not always been recognized. Piaget is well known to students of psychology and education. Usually the very simplified version of his theory is presented, and students are not introduced to Piaget's seminal text *Structuralism* (Piaget, 1970). A simple explanation for the avoidance by psychology students (and university professors) of this text is the important role played in it by group theory. Group theory is a mathematical theory. Because the mathematical education of psychologists is usually extremely limited it is difficult for them to approach this book and to grasp some of its profound philosophical insights. One of Piaget's insights concerns the way self-regulation is achieved in a structure.

To explain Piaget's thesis let me introduce briefly the key ideas of group theory. A mathematical group G is a system consisting of a set G of elements and a binary operation * on G. That is an operation on an ordered couple of elements. This system has the following properties:

1. *Closure*. The system is closed in the sense that a given operation yields only elements of the set. Formally described, for every $a, b \in G$ the operation $a * b \in G$.
2. The second property concerns the *identity* element. The fact that the group contains a neutral element e such as for each $a \in G$, $a * e = e * a = a$.
3. The third property is the existence of an *inverse* such that in combination with an element it yields the identity or the neural element. Formally, $a * a^{-1} = a^{-1} * a = e$.
4. The last property is *associativity* suggesting that for every $a, b, c \in G$, $(a * b) * c = a * (b * c)$.

Let us illustrate the notion of a group with regard to natural numbers and simple arithmetic operations. The addition of two natural numbers always results in a number that is a member of the set of natural numbers. This is an example of a systemic closure. In contrast, the operation of division does not secure closure. For example, if we divide a natural number by a bigger number then the result kicks us out of the system. One divided by two results in 0.5 and 0.5 belongs to another category of numbers. The identity of a natural number is assured by multiplying it by one.

Piaget is dealing with the idea of a structure in terms of a group. According to Piaget the self-regulation of a structure is encapsulated in the structure and there is no need for an outside observer or a meta-level activity to regulate the system. A key term for understanding this process is reversibility. Piaget (1970, p. 15) points to the fact that a binary operation is reversible in the sense that it has an inverse. For example, with regard to the arithmetic operation of addition, for each natural number n its inverse is $-n$ such as $n + (-n) = 0$ which is the identity element. Piaget argues that the reversibility results in self-regulation of the system because an erroneous result is "simply not an element of the system (if $+n - n \neq 0$ then $n \neq n$)". In other words, in a closed system, a group, the system is regulated from producing errors by its own internal logic. The idea that reversibility underlies self-regulation is a profound idea that deserves special attention, which is beyond the scope of this chapter.

Piaget noticed however that there is a crucial difference between mathematical structures and other structures whose transformations unfold in time (Piaget, 1970). In other words, there is a distinct class of structures whose transformations are governed by laws, which "are not in the strict sense 'operations', because they are not entirely reversible" (Piaget, 1970, p. 15). As Bateson has noticed too, the arrow of time differentiates biological from logical structures. The fact that time flows from past, to present, to future creates an irreversibility that is at the heart of our existence. This fact is evident to the scholar and the layman alike. My personal experience taught me an important lesson about time irreversibility. As a young soldier in the Israeli Defense Forces I had to experience basic training. During this period we were under the supervision of a sergeant major who, in trying to struggle with the basic laws of the universe, loaded us with more work than 24 hours may contain. One day he had a burst of philosophical insight. Struggling with the deadline imposed on him by his commanders, and with the need to accomplish given tasks on time, the sergeant major, stopped working, gazed at us, and declared in an authoritative yet desperate voice: "There is no mother fucker who can stop time". My sergeant major grasped in his intuitive and uneducated way a profound truth: Time flows and therefore irreversibility cannot be defeated. Piaget noticed that a group might have a heuristic value even though the transformations it involves cannot be realized physically. Moreover, he realized that structure in itself is insufficient for understanding the behavior of a system. Every structure is encapsulated in a wider explanatory frame of reference, which Piaget named the "form". At this point the similarity between Piaget and Bateson becomes evident. Every structure has a meta-structure the same as each language has a meta-language, which is a part of the language itself. Therefore, self-regulation is achieved in the interaction in between structure and meta-structure, language and meta-language. The onion-like, Klein-bottle structure of living systems is a fact rather than a philosophical obscurity.

In the first part of the book, I introduced the idea of reductionism and its shortcoming in explaining living systems. The alternative I presented framed the realm of the living as the realm of meaning making. I presented meaning as the effect of interactions produced via symbolic mediation (*i.e.* semiosis) and emphasized the contextual, inferential, and recursive-hierarchical nature of this process. I also illustrated the benefits of this perspective by offering a fresh look at a variety of biological phenomena from immune specificity to gene silencing. In the next part of the book I aim to delve more deeply into the complexity of various aspects of meaning making, from the polysemy of the sign to context, memory, and transgradience.

PART 3. ON THE WILD SIDE: FOUR LESSONS

Chapter 14 The Polysemy of the Sign: A Quantum Lesson

Summary
1 Introduction
2 Quantum Computing: The Liar and the Truth-Teller
2.1 Quantum Machines
3 From the Digital to the Analogue
3.1 Codes in Natural Language
4 Discussion

Chapter 15 Recursive-Hierarchy: A Lesson from the Tardigrade

Summary
1 Introduction
2 Recursive-Hierarchy
3 Organization Becomes Cause in Matter
4 A Recursive-Hierarchical Metabolism?
5 A Lesson from the Physics of Computation
6 From the Baron von Münchausen to the Klein Bottle

Chapter 16 Context and Memory: A Lesson from Funes the Memorious

Summary
1 Introduction: "Languaging" in Context
2 Immune Memory
3 A Lesson from Funes the Memorious

4 Immune Memory and Aging
5 Conclusion

Chapter 17 Transgradience: A Lesson from Bakhtin

Summary
1 Introduction: There is No Alibi in Existence
2 Singularity in Language
3 Bakhtin on Meaning
4 Bateson and the Mind
5 The First Law of Human Perception
6 We are All Unique but Never Alone
7 From the "I" to the Other
8 Signs as a Bridge between the "I" and the Other
9 Conclusion

Cat-logue 4

Chapter 14

The Polysemy of the Sign: A Quantum Lesson

Summary

In both natural language and biology, signs are polysemous, with a range of possible meanings before interaction-in-context determines their value. What is the meaning of *polysemy*? What is the role of polysemy in linguistic and biological systems? In this chapter, I present the idea that organisms function by using two different, orthogonal modes of communication: digital (involving discrete units) and analogue (involving continuous values). I argue that the polysemy of the sign, here metaphorically interpreted as a *superposition*, is necessary for orchestrating these modes and tying them to a concrete context of interaction.

1. Introduction

All signs are polysemous, with a range of possible meanings in different contexts. For example, in natural language the sign *shoot* can be used in one context to express an order to a soldier to fire his gun and in a different context as a synonym for speak. Polysemy is also evident in biological systems. For example, transforming growth factor (TGF) is a protein that acts as a signaling molecule between cells. An instance of TGF is TGF-B, which is described as a "*multi-functional* growth and differentiation factor responsible for regulating many diverse biological processes in both vertebrate and invertebrate species" (Zimmerman and Padgett, 2000, p. 17; emphasis mine). *Multi-functional* in this context is synonymous with polysemous; the same protein with the same structural characteristics will produce different responses in different contexts. Along the same lines, it has been argued in immunology that the meaning of a molecule being an antigen is not encapsulated in the molecule itself but emerges from the contextualization of the signal and from the communication between immunological agents (Cohen, 2000a; Neuman, 2004b). That is, the meaning of an antigen is extracted by the immune system from its position and relations in the network of communicating cells, just as the meaning of a sign in natural language is extracted by the interpreter from its position and relations in the system of signs.

To understand polysemy, and also for reasons that will be presented below, it may be helpful to discuss polysemy metaphorically as a superposition of the sign. Here the superposition of the sign is defined as *the simultaneous existence of mutually exclusive values*. Before interaction-in-context determines its meaning, a sign seems to exist in a state in which different and mutually exclusive values coexist. When I say, "The cat played the piano", the hearer can understand *cat* in several senses: the feline mammal, a jazz player, or a sexy woman. All these senses appear in the dictionary. The sign *cat* is uttered not as an isolated entity but as an integral component of a statement. When it is uttered, whether by a human being or by another system, the sign exists in between, in a state of superposition in which its different senses coexist.

The use of the term *superposition* to describe polysemy is not trivial and deserves explanation. As Nielsen and Chuang (2000) have argued:

> In most of our abstract models of the world, there is a direct correspondence between elements of the abstraction and the real world, just as an architect's plans for a building are in correspondence with the final building. The lack of this direct correspondence in quantum mechanics makes it difficult to intuit the behavior of quantum systems. (p. 13)

Something rather similar happens to us when we try to understand the polysemy of a sign that has no direct correspondence with our intuitions. Therefore, the analogical use of the term *superposition* may be justified as an instructional gambit.

At this point, an important clarification should be made. My use of the term *superposition* in the above sense does not reflect an idiosyncratic interpretation of a well-defined physical concept. I intend to examine a conceptual interpretation of superposition that is broader than the concrete, physical sense. Clearly, in the quantum sense, neither molecules nor linguistic signs can exist in a superposition. The question I am asking is whether the general conceptual sense of a superposition presented above can lead us to better understand polysemy. Therefore, this part of the book should not be considered a naïve implication of quantum computations or the application of quantum computations to phenomena that exist on a totally different scale of analysis. My use of concepts and ideas from quantum computation is for analogical purpose only. Let us digress to quantum computing in order to obtain a firm grasp of the counterintuitive work for which superposition is responsible at the subatomic level. Armed with this counterintuitive knowledge, we will hopefully be in a better position to understand the polysemy of the sign.

2. Quantum Computing: The Liar and the Truth-Teller

For instructional purposes, I would like to introduce the idea of superposition through a popular logical riddle. This step may distract us for a moment from the main argument but will give us a sense of the counterintuitive work that can be done through a superposition.

As you already realized, my conversations with my kids are sometimes a source of intellectual inspiration. A few months ago, my children challenged me with a riddle that they had heard in school. As I explained to them the solution to the riddle, it occurred to me that this riddle has a version that is insolvable in classical logic but solvable in quantum computing. Here is the riddle in its classical logical version:

> A man is standing at a crossroads. One way leads to heaven, the other to hell. The man does not know which is which. At the entrance to each road stands a guard. One of the guards always tells the truth; the other always lies. The man does not know which guard is which. He can ask a guard only one question in order to decide which road to choose. What question should he ask?

The solution is simple. The man should say to one of the guards:

> "If I had asked your colleague where this road leads, what would his answer have been?"

To find out where the road leads, the man should reverse the answer he gets, turning *hell* into *heaven* and vice versa. For example, let us assume the man encounters the liar standing at the entrance to hell. The liar knows that his colleague tells the truth and that he would have told the man that the road leads to hell. Because he is a liar, he will reverse the answer, and will say heaven instead of hell. We can see that this riddle has a simple solution within classical logic.

Let us now approach the riddle from an abstract perspective. We can think of each guard as a computing machine that takes the value of the road (*i.e.* hell or heaven) as an input and produces an output (hell or heaven). At this point it is important to make a seemingly trivial statement that was repeatedly made in this book. A computing machine does not have to look like our PC. A computing machine is just an abstract idea concerning the way an output is produced from an input.

Fig. 14.1 The liar as the logical gate NOT.

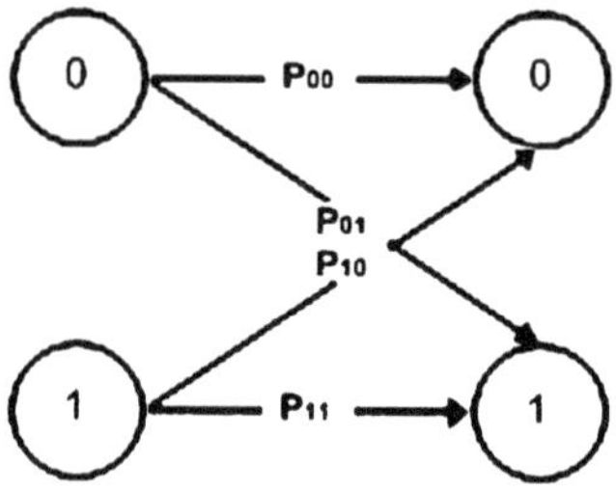

Fig. 14.2 A transition diagram from input {0, 1} to output {0, 1}.

If we represent the guards as computing machines, then the liar is a machine of *negation* that always produces the negation of the input. In other words, the liar represents the logical gate (operator) NOT (Fig. 14.1).

The truth-teller, on the other hand, is an *identity* machine that produces the output from the input by changing nothing. Our truth-teller represents the identity function that maps each value onto itself. Because NOT is the basic logical operator, identity simply means double negation—NOT (NOT(x))—which is actually copying the value from a domain into a co-domain. NOT is the constituting operation of the *digital code*, which is made up of discrete units. The digital code will be discussed later.

Now, let us complicate the riddle by assuming that each guard is a computing machine that takes the input (*i.e.* hell or heaven) and *randomly* produces an output (hell or heaven). In this version of the riddle, our unfortunate traveler seems to have no question to ask, no solution to the riddle, and no way to choose.

The reformulated riddle has no solution in classical logic. Surprising and counterintuitive as it may sound, this random version of the riddle has a solution in quantum computation.

2.1. *Quantum Machines*

Let us consider a computing machine that produces an output {0, 1} from an input {0, 1} with a certain probability (Fig. 14.2).

Let us further assume that the logic underlying the behavior of the machine is such that $P_{00} = P_{01} = P_{10} = P_{11} = 0.5$. These transformations

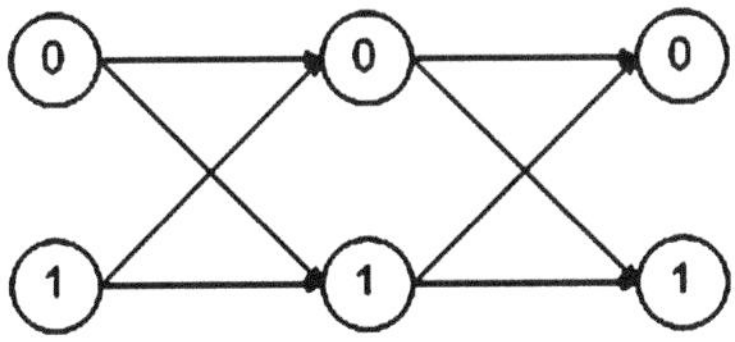

Fig. 14.3 The coupled computing machines.

represent the logical gate (*i.e.* operator) known as a coin-flip gate (CF), which randomizes its input. In other words, the probability of getting any output from any input is known and equal.

If the guards in our riddle behave like CF gates, then there is no solution to the riddle. Let us assume that the computing machines we described above are coupled and that we feed the output of the first machine into the second machine as input (Fig. 14.3).

Does it make a difference? In the case of a CF gate the answer is no, but in a case of a quantum coin-flip gate (QCF) the answer is yes! Let me explain why.

The fundamental unit in classical information theory is the *bit.* A bit of information can take only one of two mutually exclusive values, 0 or 1. In contrast, quantum computation presents a unit of information—a *qubit*—that can exist in a superposition, *being 0 and 1 at the same time.* In other words, in a qubit two mutually exclusive values coexist. Can a person be dead and alive at the same time? Can someone be a liar and a truth-teller at the same time? In contrast to the "intuition" instilled in us by classical logic, the answer to these questions is yes, at least at the quantum level. Before we approach the general conceptual meaning of a superposition, let us understand the specific meaning of the superposition of the qubit.

The superposition of a qubit—$|\varphi\rangle$—is a linear combination of the states $\alpha|0\rangle$ and $\beta|1\rangle$, where α and β are complex numbers representing the amplitudes of 0 and 1. A *probability amplitude* is a complex number-valued function that describes an uncertain or unknown quantity. In the case of the qubit, the probability of a state is equal to the square root of the absolute value of the corresponding amplitude.

In the case of our computing machine, the probability amplitude of a reflection ($0\rightarrow1$ and $1\rightarrow0$) is $1/\sqrt{2}$, and the probability amplitude of a transmission ($0\rightarrow0$ and $1\rightarrow1$) is $i/\sqrt{2}$. Figure 14.4 illustrates these amplitudes.

If we calculate the probability of getting from the input of the first machine to the output of the second machine, we find that the probability of

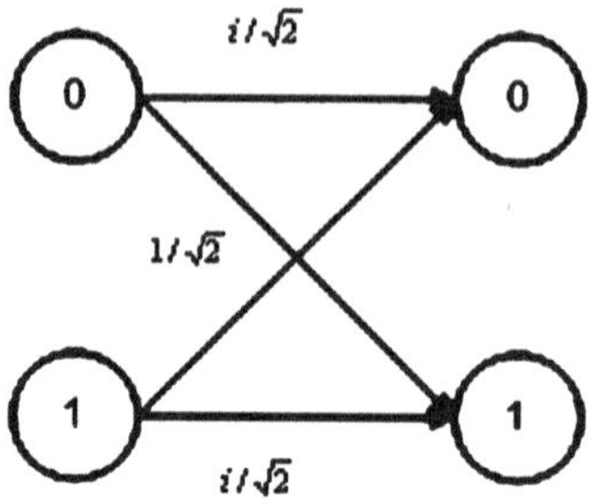

Fig. 14.4 Probability amplitudes of reflection and transmission.

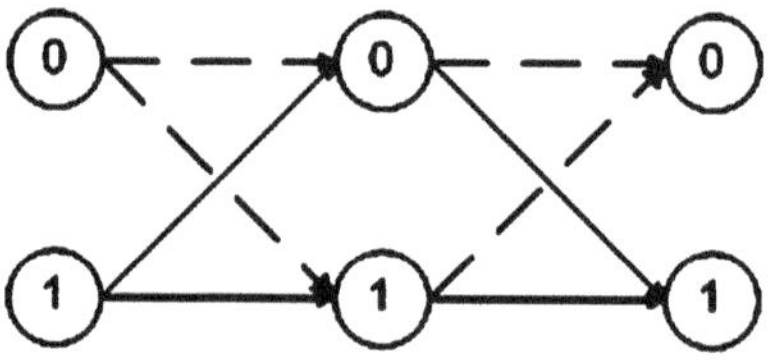

Fig. 14.5 The path from 0 to 0.

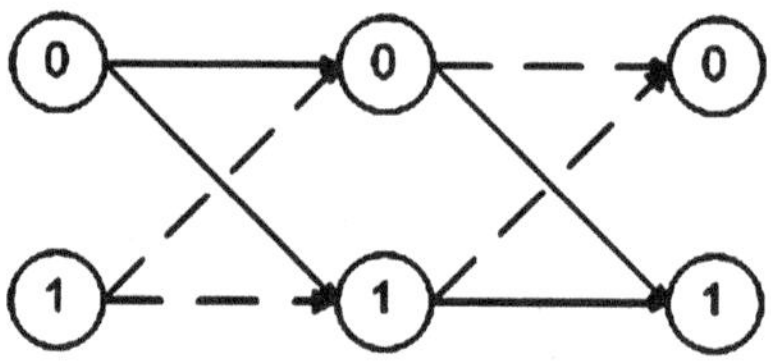

Fig. 14.6 The path from 1 to 0.

getting from 0 to 0 (here represented by dashed lines in Fig. 14.5) is: $P_{00} = |(i/\sqrt{2})(i/\sqrt{2}) + (1/\sqrt{2})(1/\sqrt{2})|^2 = 0$

In a similar way we find that the path from 1 to 0 is (Fig. 14.6):

The transition probability is

$$P_{10} = |(i/\sqrt{2})(1/\sqrt{2}) + (1/\sqrt{2})(i/\sqrt{2})|^2 = 1$$

The analysis of the two successive operations shows that when the input into the first gate is 0, the output of the second gate is 1, and when the input into the first gate is 1, the output of the second gate is 0.

It turns out that when the QCF machine is followed by an *identical* machine, the final output is always *the negation of the first input*. In other words, two successive QCF gates implement the NOT function. If

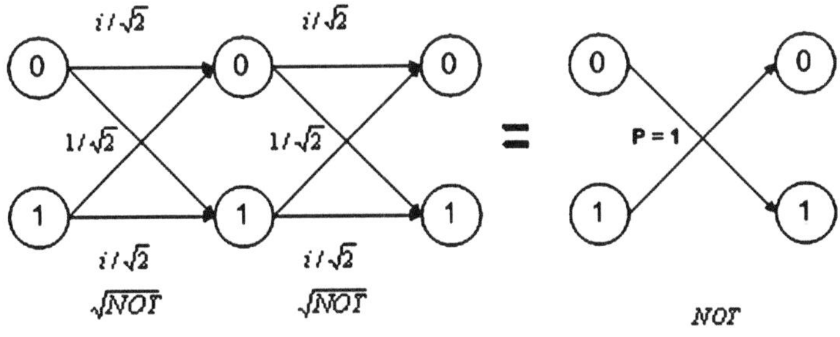

Fig. 14.7 The square root of NOT.

$(QCF)^2 = NOT$, then a single QCF gate can be said to calculate the "square root of NOT" (Brown, 2000). Figure 14.7 illustrates this process.

The conclusion sounds counterintuitive. How is it possible that a machine produces an output of 0 or 1 with equal probability and that two identical machines, acting independently, produce a deterministic output?

Surprisingly, such machines exist in nature, albeit at the quantum level (Brown, 2000; Deutsch *et al.*, 2000). Their bizarre effect becomes clearer if we realize that the two transitions that make up the flow from the input to the output occur *simultaneously* and that they cancel each other out through *quantum interference*.[1] In other words, when the amplitudes of the quantum states have different signs, they cancel each other out and the interference is destructive.

Let us explain this phenomenon better by showing how a superposition combined with the square root of NOT produces this counterintuitive outcome. First, it is important to note that the input of the first machine is in a state of superposition. If the input is 0, then it comes out as 0 with an amplitude of $1/\sqrt{2}$; if it is 1, then it comes out as 1 with an amplitude of $i/\sqrt{2}$. Let us describe the base states as:

$$\begin{pmatrix}1\\0\end{pmatrix} \equiv |0\rangle, \quad \begin{pmatrix}0\\1\end{pmatrix} \equiv |1\rangle$$

The logical gate of the square root of NOT is the operator:

$$1/2\begin{vmatrix}1+i & 1-i\\ 1-i & 1+i\end{vmatrix}$$

[1] The superposition of two or more waves resulting in a new wave pattern.

If we start with 0 and apply the square root of NOT, and then apply the same operator to the product, the output is 1. If we start with 1 and run the same procedure, the final output is 0.

Now, we understand the work that can be done with a superposition and we see how the riddle we presented before is solvable in its quantum version. If each guard is a quantum machine, then we simply couple the machines and use the first guard's answer as the input for the second guard. The output of the second guard should be reversed in order to discover at which road the first guard is stationed.

The important lesson to draw from the above discussion is that a superposition can allow us to conduct logical operations that have no equivalence in classical logic and no basis in our intuition.

Now that we realize the unique work that a superposition can do at the quantum level, we may ask what work a superposition can allow at the semiotic level. In other words, assuming that a sign, the semiotic particle, can hold several mutually exclusive values, what can be explained by this phenomenon?

3. From the Digital to the Analogue

To explain the work polysemy/superposition can do, let us turn again to Gregory Bateson and the idea that the basic unit of information is a "difference that makes a difference". Let us start with a difference before turning to a difference that makes a difference. A difference is a *qualitative* relation between at least two entities, and the most basic relation is that between the binary categories 1 and 0, in which each entity is defined simply as the negation of the other. That is, the basic operation of the difference system is the logical gate NOT and each unit is defined only in *negative* terms. There is no positive or essential meaning that can be assigned to 1 or 0. In this sense, the system of differences corresponds to classical logic.

The minimum difference involves a two-digit string, but it can be extended to a finite string, which is actually the tape of a Turing machine. Bateson describes this string as the digital code or, more accurately, as the digital mode of communication. The terms *digital code* and *digital mode* will be used interchangeably. The digital mode and its combinatorial capacity underlie the evolutionary flexibility of the living system (and natural language) and its potential to create novelties from discrete units such as genes or signs.

The digital mode has several characteristics indispensable for understanding life (for a discussion, see Hoffmeyer and Emmeche, 1991). The most important characteristic of the digital code is its arbitrariness. The digital mode of communication is *arbitrary* and is not bound strictly to the message it carries. In other words, there is no *logical necessity* that a given sign will correspond to a given signified. This arbitrariness is true of any system of digital codes, whether organic codes or codes in natural language. For example, there is no logical necessity that the codon CUU code for leucine. As emphasized by Barbieri (2004), convention is at the heart of the digital code.

In natural language the arbitrariness of a sign means that the correspondence between the sign and its conceptual content (*i.e.* the signified) is not obligatory but emerges from the interactions between members of a given society. The fact that a cat is *cat* in English and *gato* in Spanish is just one instance of this arbitrariness and its conventional nature.

3.1. Codes in Natural Language

The idea that a difference is our basic unit of analysis also appears in the seminal work of Ferdinand de Saussure with respect to another digital system—natural language. To review, for Saussure, "In the language itself, there are only differences. Even more important is the fact that, although in general a difference presupposes positive terms between which the difference holds, in language there are only differences and no positive terms" (Saussure, 1972, p. 118). Because a sign is arbitrarily related to its signified, it is impossible to define a sign in positive terms. Each sign is negatively defined as being different from the others. The fact that the logical operator NOT is the constituting relation of the digital mode explains this argument. A sign has no defining essence. It is an arbitrary entity that is defined as not being another sign. In another hypothetical culture, the sign cat could potentially stand for the member of the genus Canis, and dog could stand for the feline mammal. It is important to understand the context of this suggestion. For Saussure, language as an abstract system of signs (la langue) is "a system of distinct signs corresponding to distinct ideas" (Saussure, 1972, p. 26). That is, in itself a sign means nothing. It exists solely by being differentiated. According to this interpretation the sign cat has no intrinsic meaning. The catness of the cat is not embedded either in the way the word cat is pronounced or in concept of a cat. The meaning of the word cat emerges from its position and relations in the system of signs.

Saussure's argument concerns the sign as an isolated unit that is "purely differential and negative" (Saussure, 1972, p. 118) as a phonetic or conceptual unit. However, there is another dimension of the natural sign system. It should be recalled that for Saussure the meaning of a word is the conceptual "counterpart of a sound pattern" (Saussure, 1972, p. 112). In this sense the meaning of the sign *cat* is its corresponding concept of a cat. Saussure suggests that *meaning* should be distinguished from *value*, which is important for understanding the abstract nature of any system of signs. If difference is the first dimension of the sign system, *value is the second.*

A semiotic value is an abstract concept that, according to Saussure, involves "(1) something *dissimilar* which can be exchanged for the item whose value is under consideration [*e.g.* a $1 bill for ice cream or a sign for a signified] and (2) *similar* things which can be *compared* with the item whose value is under consideration" (*e.g.* a word for a synonym) (Saussure, 1972, p. 113). Like the monetary system, natural language is constituted through exchange and value. If we think about a semiotic system in terms of a network, then differences are expressed as the different parts of the network (*i.e.* the nodes) and value is expressed in terms of their relative positioning.

A value belongs to another mode of communication—the analogical mode—that involves continuity. Similarity and dissimilarity are a matter of degree. Therefore, the analogue code concerns relations of magnitude and has no signal for *not* (Bateson, 2000, p. 291). While NOT is the constituting operation of the digital code, it does not exist in the analogue code. The analogical mode is the realm of the continuous. The realm of the analogue code is not the realm of classical logic. It is the realm of the Freudian unconscious, the realm of symmetric (Matte-Blanco, 1988), and fuzzy logic. It is the realm in which anything is connected to anything to a certain degree.

The language system, like any other semiotic system, is a system of pure values whose function is to combine *two orders of difference*: digital and analogue, difference and value. The two modes are crucial for understanding any system of signs in which differences, negatively defined, have to take on values in order to make a difference that makes a difference. Pure differences mean nothing if they do not have a value. For example, DNA as a string of differentiated letters is informative only for a given interpretative system, and only if it can make a difference for the production of proteins.

Any semiotic system combines the two orders of difference: the digital and the analogue. Language, however, like any other semiotic system, is more than an abstract system of differences and values. In natural language a difference makes a difference only for someone (or something) in a concrete context of interaction. Language exists in speech—*parole*—in a concrete context of interaction and exchange. Remember Volosinov? This point was made clear by Volosinov (1986), when he criticized Saussure for emphasizing the abstract *langue* over the concrete *parole*. If we understand this point, then we cannot accept the sharp Saussurean division between *langue* and *parole*, and we cannot accept the desperate Saussurean withdrawal from the complexity of *parole* to the simplicity of the abstract *langue*.

Language, or any semiotic activity, is a wholeness constituted by the delicate interplay between the abstract and the concrete, *langue* and *parole*. This point is beautifully illustrated by a poem written by the Nobel laureate Wislawa Szymborska. In her beautiful poem "Sky" she writes:

> Even the highest mountains
>
> Are no closer to the sky
>
> Than the deepest valleys.
>
> There's no more of it in one place
>
> Than another.
>
> And
>
> Division into sky and earth
>
> It's not the proper way
>
> To contemplate this wholeness.

The problem facing us is understanding how living systems "contemplate this wholeness", how people, for example, stitch the abstractness of *langue* to the concreteness of *parole*? How does the immune system usually treat sperm cells as "non-self" but in a specific context as self? The answer I would like to provide concerns the polysemy of the sign.

The realm of *langue* is the realm of differences and values. This realm is constituted and maintained by convention, social or biological, which is characterized by arbitrariness. Arbitrariness underlies convention, and convention underlies flexibility and resilience. However, convention and arbitrariness have their price, namely, that they cannot directly support the generation of meaning in context. Direct translation from the abstract to the concrete is impossible. The realm of differences and values is the realm of pure potentialities. It is the "post-modern" realm in which anything goes. It is the nightmare of the interpreter. On the other hand, the realm of concrete interaction is the realm of actualities in which a sign should signify a specific sense. How do we bridge the gap between sky and earth? Between *langue* and *parole*? Between the abstract and the concrete? Between the potential and the actual? Between Plato's forms and Heraclites' river? My answer is through the polysemy/superposition of the sign.

The superposition of the sign is a state in which we "lift" the sign from its embodiment in an abstract system of relations and positioning. If we let its potential values coexist, the sign is no longer in the realm of pure, mutually exclusive potentialities. On the other hand, it is not in the realm of actualities either. It exists in between. Only by paradoxically existing in between does the sign allow us to bridge the gap between the abstractness of *langue* and the concreteness of *parole*.

There is another aspect to the arbitrariness of the sign, and this aspect concerns its irreversibility. To review, an irreversible process is a computational process in which the input cannot be reproduced from the output. For example, I have in my mind the concept of a cat. When this concept is translated into natural language, it is represented by the particularities of a given language, such as *cat* in English or *gato* in Spanish. However, the sign itself is irreversible, because the sign *cat* can be used in different senses, such as slang for a sexy women or a jazz musician. That is to say, from the sign itself (*i.e.* the output) I cannot recover its conceptual counterpart (*i.e.* its input). Again, convention and arbitrariness have their price. In this sense, the sign is the product of an irreversible process. Only context determines the sense of the sign and allows us to restore its meaning. Counterintuitive as it may sound, the superposition of the sign puts us in a situation in which the original input can be restored despite having necessarily been subjected to an irreversible process of computation (Neuman, 2006), as implied by the arbitrariness of the coding system.

4. Discussion

To conclude, let us turn to Borges and his short story "The Garden of Forking Paths" (Borges, 1962). In this story Borges discusses the idea of a *chaotic novel*. He says: "In all fiction, when a man is faced with alternatives he chooses one at the expense of the others".

But in the chaotic novel he chooses "simultaneously—all of them. He thus *creates* various futures, various times which start others that will in their turn branch out and bifurcate in other times" (p. 98).

The chaotic novel is similar to the superposition of a subatomic particle or a sign. According to this interpretation, a sign exists in a slice of a multiverse—several coexisting universes of discourse. The sign has degrees of freedom in the sense that it may mean different things in different contexts. Each sign has a counterpart in a range of other universes, and its actual value is determined in interaction-in-context. In the chaotic novel the man *simultaneously* chooses all of his alternatives instead of just one of them, whereas in a concrete interaction a choice has to be made.

Borges discusses the garden as a metaphysical model:

> "The Garden of Forking Paths" is not just a novel. It is a "picture, incomplete yet not false, of the universe such as Ts'ui Pen [the author of the novel] conceived it to be. Differing from Newton and Schopenhauer, your ancestor did not think of time as absolute and uniform. He believed in an infinite series of times, in a dizzily growing ever spreading network of diverging, converging and parallel times". This web of time—the strands of which approach one another, bifurcate, intersect or ignore each other through the centuries—embraces *every* possibility. (Borges, 1962, p. 100)

May be we live in a "garden of forking paths", as some physicists think, but if we live in a realm in which our life does not have a counterpart in other universes, then the superposition of the sign is just a practical and mediatory step that allows an organism to function by maintaining a delicate balance that constitutes its life here and now. In this sense, and in contrast with the superposition of a physical particle, the superposition of the sign is not an *ontological* state but an *epistemological* stance. Living systems behave as if the sign were in superposition. This is an epistemological stance, and as Gregory Bateson taught us more than once, some epistemological stances pay off.

In this chapter, I delved more deeply into the meaning of polysemy. I explained polysemy in terms of a superposition and showed the work it can do in materializing the logic of in between. Let us close this chapter by linking it to a previous discussion. Do you remember Maxwell's demon? Remember the intelligent creature that decides between two options? Isn't the demon, at its most basic level, a measurement process that turns an indeterminate signal into a difference that makes a difference?

Chapter 15

Recursive-Hierarchy: A Lesson from the Tardigrade

Summary

The tardigrade is a small microscopic creature that under environmental stress conditions undergoes cryptobiosis, a temporary metabolic depression, which is a third state between life and death. Cryptobiosis is an unexplained phenomenon. It is argued that this state is biologically obscure from a biological reductionist point of view, however, cryptobiosis makes sense within a different theoretical framework. The ability of the tardigrade to bootstrap itself is interpreted according to Gregory Bateson's idea of a *recursive-hierarchy* and a topological perspective on thermodynamics. It is argued that the structure of the organism is a recursive-hierarchical structure that allows the organism to conduct processes of reversible computations of which cryptobiosis is just a specific instance. The general meaning of this conclusion is discussed in the context of a scientific non-reductionist approach to biological systems and is used to illustrate the notion of a recursive-hierarchical system.

1. Introduction

One day, I sat with my kids and watched some kind of a naturalistic program on television. An amazing creature was introduced to the audience. A small microscopic creature (250–500 μm) that is known as the "water bear" or as "Tardigrade, the slow walker". This microscopic creature is hardly known either to biologists or to laymen, although it can be found almost everywhere on earth from the top of the Himalayas to the bottom of the oceans. This microscopic creature is named the "water bear" because the first person that published a paper about it in 1773 described it as resembling a bear in miniature. Figure 15.1 is a painting of the water bear and you can see why it got its name.

The tardigrade has its own phylum *Tardigrade* and its unique characteristics (Nelson, 2002): a thick cylindrical bilateral symmetrical body with four segments and a head with eyes, four pair of legs, feet with claws or toes, ventral nervous system, and a multilobed brain.

Fig. 15.1 The tardigrade.

The tardigrade is known for its ability to survive in extreme environments including complete dehydration, boiling water at 151°C, and an amount of radiation that is thousands of times stronger than the amount of radiation that is lethal to human beings.

Organisms have different mechanisms for adjusting to environmental stress conditions. The tardigrade is described in this chapter because it is capable of entering a latent state—*cryptobiosis*—when environmental conditions are unfavorable (Nelson, 2002). David Keilin coined the term "cryptobiosis"—hidden life—and defined it as:

> the state of an organism when it shows *no visible signs of life* and when its metabolic activity becomes hardly measurable, or comes reversibly to a standstill. (Keilin, 1959; quoted in Clegg, 2001, pp. 6, 13; emphasis mine)

Feofilova (2003) and other researchers argue that since the metabolic activity of the organism is hardly measurable it is completely inhibited. However, it is possible that during cryptobiosis metabolism exists at a very low level that is not detectable by measurement procedures. This possibility will be explored in this chapter. However, modern scientific knowledge is based on the positive products of measurement procedures and not on speculations resulting from the limits of measurement procedures. Therefore, we should accept, at least as a starting point, the common knowledge of the field suggesting that when a tardigrade is in a latent state of cryptobiosis "metabolism, growth, reproduction, and senescence are reduced or cease temporarily"

(Nelson, 2002, p. 655). In fact, it is argued that cryptobiosis involves a "complete or a near-complete inhibition of metabolic activity (0%)" (Feofilova, 2003, p. 2). Since metabolism is a defining characteristic of life one can argue that cryptobiosis is a kind of *temporary death.*

Beyond the quantitative metabolic aspect of cryptobiosis we should realize the qualitative aspect of cryptobiosis. Each organism finally dies and death is clearly an irreversible state. To exclude miraculous stories, such as those described in the Bible and the New Testament, a dead organism cannot be revived. Death is irreversible and it is the final stage of the organism's journey along the arrow of time. However, an organism that is in a state of cryptobiosis is in a unique state that is somehow a state of a *potentially reversible death*. Indeed, due to its reversibility, cryptobiosis is considered to be a unique biological state between life and death (Clegg, 2001).

The depression of metabolism in the face of environmental stress is acknowledged as "a normal part of the life cycle of many animals, and it has been reported in most of the major invertebrate's phyla and in all vertebrate classes" (Guppy, 2004, p. 435). The ubiquity of cryptobiosis may lead us to expect a wealth of knowledge about the mechanisms underlying cryptobiosis. Surprisingly, there are only 32 references to cryptobiosis in the PubMed (August 2005). This state probably indicates the theoretical obscurity of cryptobiosis and the fact that it is poorly understood.

This conclusion is supported by biologists who study cryptobiosis. As argued by Wright (2001, p. 564): "Tardigrade cryptobiosis remains poorly understood". Schill *et al.* (2004) argue that cryptobiosis in tardigrades and other invertebrates is characterized by several major events that "still remain largely unidentified" (p. 1607), and Watanabe *et al.* (2002) argue that the underlying molecular and metabolic mechanisms of cryptobiosis largely remain a mystery. Concerning metabolic depression it was recently argued by Guppy (2004, p. 436) that "to date, no molecular mechanism or process associated with the control of metabolic depression has been comprehensively delineated, and the fundamental phenomenon of metabolic depression remains biochemically obscure".

The tardigrade, although not showing (or hardly showing) any signs of life at the metabolic level, is capable of "reviving" itself by responding to cues of a friendly environment. In this case, a drop of water is enough to signal the tardigrade that it can come back to life. The organism must respond to this signal, and make sense out of it. Here we encounter a difficulty. If we accept the argument that cryptobiosis is a unique biological state in between life and death, how is it possible for the organism to interpret the signal? How is it possible for the tardigrade to extract itself from a state in which metabolism does not exist or almost does not exist? How can it move from a state in which no free energy is allegedly available to maintain biological functions?

The mystery of cryptobiosis may be attributed to the lack of appropriate conceptual tools for approaching these questions. The following sections aim to point at possible directions of inquiry. More specifically, I speculate that (1) cryptobiosis involves a shift toward a form of reversible computation and (2) the bootstrapping from an irreversible computation to a regular reversible computation takes place in the topology of a recursive-hierarchy. These speculations are valuable for several reasons. First, they are speculations that may turn into hypotheses that will direct empirical research. Second, they may challenge theoretical biologists by initiating a discussion about the meaning of a bootstrapping process in living systems. Third, they may show that, as Strohman (2000) argued, "organization becomes cause in matter" and that this kind of organization should be the subject of a scientific inquiry.

2. Recursive-Hierarchy

One of Bateson's most fruitful ideas, later described as a recursive-hierarchy (Harries-Jones, 1995; Neuman, 2004b), was that all living systems are multilevel and recursive. For example, Bateson noticed that informational content in biological systems always assumes a context of interpretation where the term context is used in the sense of a higher-order form or constraint. The context is the one that restrains (to use Bateson's terminology) or constrains (to use a terminology, I have used elsewhere; Neuman, 2004b) the entropy of the system and its natural tendency toward disorder.

The abstract idea of constraints may be illustrated through the simple mechanical example of Brownian ratchets (Dill and Bromberg, 2003, p. 330). The example concerns the way molecular machines produce directed motion. Brownian motion in itself cannot be used for a directed motion because it is random. However, the Brownian ratchet model suggests how random diffusion coupled with energy-driven, but non-directional binding and release events, can lead to directed motion. Let us consider a molecule M that moves itself along molecule P having a chain of binding sites (Fig. 15.2).

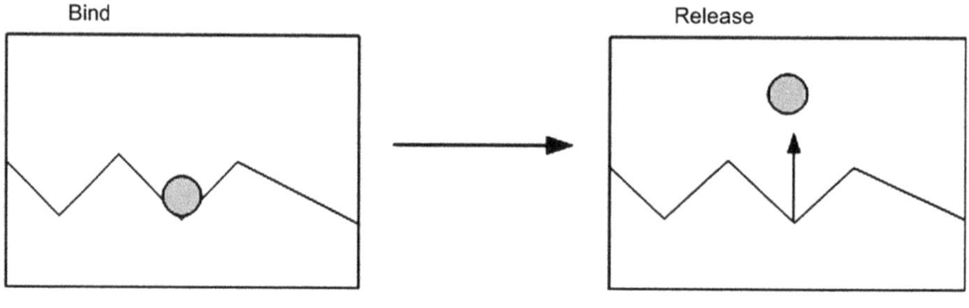

Fig. 15.2 The ratchet model.

Before time $t=0$ the system is stable and M is bound at a location where the binding free energy—$F(x)$—is at a minimum. At time $t=0$ energy is put in the system to release M (signified in the picture by a ball) from its binding site on P (its landscape). M is free to move either in the $+x$ or $-x$ direction along P. M remains free and diffuses. This diffusion leads to a Gaussian distribution along x. During that time some of the molecules will diffuse to $x \geq a$, where a is used to denote the location of the next maximum to the right. Those molecules will rebind and slide energetically downhill to the next energy well to the right of the binding site. A smaller number of molecules will diffuse to the left. The diffusion is symmetrical in x. However, the ligand-binding potential is not. The free energy is asymmetric. At the time of the diffusion more particles fall into the energy well to the right (or more accurately to a direction which is determined by the asymmetry) of the binding site. Therefore, if there is an appropriate time lag between the cycles, repeated cycles of release, diffusion, and rebinding will lead to the directed movement of M.

Two things should be emphasized. First, the Brownian ratchet model does not violate the second law of thermodynamics. Second, and much more relevant for our case, the molecule M in the ratchet model is not directed in a given direction by external forces or by its own erratic movement. The directional movement of the molecule is achieved by imposing energetic constraints on the Brownian movement. This is a simple mechanical example illustrating the way higher-order constraints of the system "determine" the behavior of a lower-level entity.

Context is a "collective term for all those events which tell the organism among what *set* of alternatives he must make his next choice" (Bateson, 2000, p. 289). Rather than allowing Brownian erratic movement to control the system's behavior, the context as a higher form of organization directs the system's trajectory toward a given attractor. This idea was found to be fruitful in explaining processes at different scales of analysis. For example, it was used to explain immune recognition as a meaning-making process (Neuman, 2004b, 2005), and the way meaning is related to information (Neuman, 2006).

A context is always embedded within another context and therefore we have a *hierarchy of* constraints *of* constraints. The fact that contexts are embedded within contexts does not point only at the hierarchical structure of living systems. There is also a dynamic aspect to this embedding and embodiment, and this dynamic aspect is constituted through feedback loops in which information is fed back and forth between the different levels of the system to assure the stability of each level and to constitute the working whole. Therefore, organisms can be described as *recursive-hierarchical* systems (Harries-Jones, 1995; Neuman, 2004b) that are capable of self-determination through embedded levels of constraints constituted by feedback loops. The important implication of this statement is that a system of a recursive-hierarchy is a system capable of *self-determination* and therefore has the potential of bootstrapping. As already envisioned by Bateson (2000):

> If, in the communicational and organizational processes of biological evolution, there is something like *levels*—items, patterns and possibly patterns of patterns—then it is logically possible for the evolutionary system to make something like positive choices. (p. 411)

Notice that Bateson uses the expression *positive choices* not in the intentional sense but in the sense of self-determination of higher levels on the behavior of a lower level.

Bateson's idea of a recursive-hierarchy should not be confused with the idea of strong downward causation in which "a given entity or process on a given level may causally inflict changes or effects on entities or processes on a lower level" (Emmeche *et al.*, 2000, p. 19). The idea of strong downward causation, in which a higher-order level determines the behavior of a lower level, is problematic and incompatible with our knowledge of physics. Bateson's idea is closer to the idea of *medium downward causation* in which "an entity on a higher level comes into being through a realization of one amongst several possible states on the lower level with the previous states of the higher level as a factor of selection" (Emmeche *et al.*, 2000, p. 24) meaning that the higher level serves as a *boundary condition* or constraint *condition* on the behavior of the lower level. That is the higher level is characterized by "organizational principles" (Emmeche *et al.*, 2000, p. 25) that have an effect on the distribution of lower-level events and substances. For example, although fluctuations appear on the quantum level, stability appears on the molecular level due to the constraints imposed by the molecular organization on the atomic and subatomic degrees of freedom. This idea is highly relevant for the tardigrade case. It implies that although micro-level metabolic activity may be significantly reduced, high-level forms of organization may still be maintained to allow at the right moment a shift to normal metabolic activity.

After illustrating the meaning of constraints, it is the time to explain the meaning of a bootstrapping procedure. The term *bootstrapping* alludes to the legendary Baron Münchausen who was able to lift himself out of a swamp by pulling up on his own hair or his own bootstraps (http://en.wikipedia.org/wiki/Bootstrapping). The term acquired different senses in a variety of domains. For example, in computer science, this term refers to any process where a simple system activates a more complicated system. It is the problem of starting a certain system without the system already functioning, a process that may be portrayed as allegedly illogical or paradoxical the same as the

Baron's legend. However, solutions, accordingly called bootstrapping, exist; they are processes whereby a complex system emerges by starting simply and, bit by bit, develops more complex capabilities on top of the simpler ones. The tardigrade's shift from normal activity to cryptobiosis may be portrayed as a logical biological process of slowing down metabolic processes. However, a shift from a state of cryptobiosis to a state of normal metabolic activity is a bootstrapping procedure. It is not simply a shift from a lower level of metabolic activity to a higher level, but a qualitative shift from one state to another more complicated state that involves higher-order behaviors such as reproduction and predation. The next section explains the possible logic underlying the tardigrade's bootstrapping process.

3. Organization Becomes Cause in Matter

Bateson's idea of a recursive-hierarchy might be misinterpreted as an expression of a general non-scientific holism. However, this idea has been fully realized in modern conceptions that consider organization as cause. In 2000, Richard Strohman published an insightful commentary in *Nature Biotechnology* entitled: "Organization becomes cause in the matter". Strohman's commentary may be used to apply Bateson's ideas to the tardigrade and the mystery of cryptobiosis.

Strohman's point of departure is the perspective of complex systems and "the middle way" (Laughlin *et al.*, 2000) or the search for the laws operating at levels and scales of organization intermediate between the microscopic state of fundamental particles and the macroscopic state of higher-level organization. Following Laughlin *et al.* (2000), Strohman argues that

> in biology, molecular genetic reductionism has mostly distracted us from study of mesoscopic realms between genotype and phenotype where complex organizational states exists, and where, as we now realize, there also exist networks of regulatory proteins capable of reorganizing patterns of gene expression, and much other "emergent" cellular behavior, in a context-dependent manner. (Strohman, 2000, p. 575)

Strohman argues further that it is the mesoscopic level which is responsible for the emergent features of biological systems, and pays his intellectual debt to the work of Michael Polanyi, whose idea of boundary condition (Polanyi, 1968) is in line with Bateson's idea of a recursive-hierarchy, and the idea of medium downward causation. A similar idea also appears in Yates (1993) who argues that "every 'level' in a natural system is constrained by the next level below and the next level above; it is in a middle of a sandwich, with every level equally sovereign with respect to the global stability of the organism. Command, control, communication, and co-operatively permeate 'laterally' (heterarchically) and 'vertically' (bottom-up and top-down)" (Yates, 1993, p. 212). Strohman also points to *metabolic* networks as an arena in which the new venture of mesoscopic analysis may show its power. This idea brings us back to the tardigrade and the way it depresses its metabolism. However, before delving into this point, let me quote an excerpt from a manuscript written by a leading experimental immunologist. This manuscript, which is a theoretical treatise in both immunology and theoretical biology, deals with life as an emergent property (Cohen, 2000a). This statement is worth reading because it reminds us that life is in the organization and that in a very deep sense, organization is the cause of living matter.

> The clearest example of an emergent property is life itself. Life is not inherent in any single element constituting the living cell. DNA is not alive, neither are proteins, carbohydrates or lipids. Indeed, for a single short moment, a living cell and a dead cell may, upon analysis, be found to contain precisely the same catalogue of "dead" chemicals in identical concentrations. Bacteria have been resurrected after 35 million years of suspended life in the guts of ancient bees entrapped in amber. While not quite dinosaurs, 35 million year old bacteria are still a marvel. Today they surely live; what was their state for 35 million years? What distinguishes the living from the dead? Nothing more than *actions and interactions*. Life emerges from inert matter as a consequence of metabolism, the continuous transfer of energy and information systematically packaged in cells in a way that leads to self-perpetuation. (Cohen, 2000a, §47)

Cohen's statement naturally leads us to the next section.

4. A Recursive-Hierarchical Metabolism?

Metabolism involves the "autonomous use of matter and energy in building, growing, developing, and maintaining the bodily fabric of a living thing" (Boden, 1999, p. 237). Bergareche and Ruiz-Mirazo (1999) define metabolism in the most abstract sense as

> any material organizational apparatus of energy management which can implement an operationally close constructive-relational system, so that the network of components production relations held in it *recursively* maintains and renews the aforementioned apparatus. (p. 53; emphasis mine)

This abstract definition of metabolism suggests that *metabolism operates as a recursive-hierarchical structure*, which as we discussed before, has the potential of bootstrapping through a medium version of downward causation. If during cryptobiosis the tardigrade is capable to maintain a minimal level of metabolic activity, just enough for maintaining the organization of the metabolic network, then this organization (*i.e.* a higher level of the metabolic network) may serve to return the metabolic activity of the cells to a normal level. This suggestion assumes the possibility of constituting minimal maintenance activity in the network at minimal energy expense, and a unique topology of the metabolic network that materialize the recursive-hierarchical structure. It is not quite clear what the meaning of metabolic constraints is at a higher level. A possible interpretation is that while the energy-consuming activity of translation from DNA is depressed, a minimal level of energy is used for maintaining the organization of network in itself, maybe through the mitochondria that has its own DNA. A more biologically oriented explanation is currently not at hand and may be the aim of a future work. The next sections point to the viability of this possibility in the context of the physics of computation.

5. A Lesson from the Physics of Computation

To review, Landauer and Bennett (1985) describe a process of computation, in the most general sense, as a process in which an *output* is produced from an *input* and *information* is considered in the most general sense of differentiated states. They argue that this process of computation has a clear physical meaning. Processes of computation do not take place in a Platonic space of ideas but are physically grounded. This perspective led them to offer a thermodynamic approach to computation. According to this approach, "the digital computer may be thought of as an engine that dissipates energy in order to perform mathematical work" (Bennett, 1982, p. 906). The approach has some interesting insights with clear implications for biology.

For example, one of the insights of the physics of computation concerns the price of eliminating information (to include biological information) from the system. Landauer (1961) argued that the elimination of information from a given system is an activity that consumes energy and dissipates heat into the environment. "When an information is erased there is always an energy cost of $kT \ln 2$ per classical bit to be paid" and "amount of heat equal to $kT \ln 2$ is dumped in the environment at the end of the process" (Plenio and Vitelli, 2001, p. 27).

The physics of computation suggests that computation, which is usually discussed in a purely functional way, is a physically grounded process, which is subject to the laws of thermodynamics. Landauer and Bennett (1985) push this idea forward to biological systems, and we can argue that the organism is a unique type of highly complex metabolic engine that dissipates energy to perform "biological work". In this context, the difference between *reversible* and *irreversible* processes is of great importance.

In thermodynamics, a reversible process changes the state of a system in such a way that the net change in the combined entropy of the system and its surroundings is zero. It is a process in which no heat is lost from the system as "waste" and the machine is as efficient as it can possibly be. In other words, the process does not result in the increase in physical entropy and the loss of information.

A reversible computing is a computational process that is reversible at least to some close approximation, and it has the merit of improving the energy efficiency of the computer or the organism using it. *It must be emphasized that the idea of reversible computation does not contradict the second law but just questions the limits of the price organisms (and other computational devices) should pay* for it as open systems. Indeed, Bennett presents some theoretical models that perform computation with (approx.) zero energy dissipation. Although these theoretical models have not, as yet, been materialized by human beings, there is no theoretical obstacle to their existence at the molecular level during cryptobiosis.

It is possible to imagine a reversible process of computation that strives for a minimal level of energy expenditure. It is also possible that cryptobiosis involves such a process of reversible computation. This speculation is grounded in the slowdown of all metabolic processes during cryptobiosis. This slowdown may reflect a shift to reversible computing, which is capable of maintaining a very low level of metabolic activity with minimal energy expenditure. This activity may maintain the organization, which at the right time will support the bootstrapping procedure, the emergence of a higher-order state, and the shift to normal metabolism.

Computation as it is commonly materialized in the electronic computer is irreversible and the dissipation of energy and the loss of information are inevitable. Cognitive systems also involve irreversible processes of computation in which micro-level differentiations are lost in favor of higher-level outputs. For example, seeing involves an irreversible process of computation in which lower-level changes in the activity of retinal cells is integrated at a higher level and produce the perceived image. Irreversibility is a defining property of a hierarchical system. When we shift from one level of organization to a higher level of organization, a certain level of differentiation and information is, by definition, lost in the transition due to energy dissipation. However, in the case of a recursive-hierarchical structure this does not necessarily have to be the case, and the dissipation may be reduced to the limit line.

In sum, if the metabolic network of the tardigrade is built according to the topology of recursive-hierarchy, and if it shifts to a reversible process of computation during cryptobiosis, then metabolic activity would be undetectable (or barely detectable) although it will be efficient enough to bootstrap the organism.

If cryptobiosis works according to the logic of reversible computation, which is materialized in the recursive-hierarchical structure of the metabolic network, then the ability of the tardigrade to bootstrap itself turns from a mystery into a process that is comprehensible in scientific terms.

We may conclude our analysis so far by suggesting that while death is the ultimate expression of an irreversible process called life, cryptobiosis is a state in between life and death involving a temporary shift to a reversible form of biological computation. The organism is able to bootstrap itself when macro-level constraints reorganize and allow the highly efficient metabolic computation to re-use resources of energy from the environment.

6. From the Baron von Münchausen to the Klein Bottle

Our scientific knowledge is mediated and impeded by the models we use, including visual models of representation. For example, according to Darwinian theory, the environment is portrayed as a kind of a *strainer*. Random mutations of DNA result in different phenotypes that pass or do not pass through the strainer. The general epigenetic conception puts much more weight on the shoulders of the environment. The environment is not portrayed as a strainer but as a *landscape* in which the potentialities of the organism are channeled. This metaphor was introduced by Waddington (1957), whose visual image of an epigenetic landscape has become well known. Figure 15.3 shows an epigenetic landscape as it was presented in Waddington's *The Strategy of the Genes*.

The developing embryo, according to Waddington, is like a ball channeled by the structure of the landscape. This portrait does not give the organism, whether at the genetic, embryo, or mature level, any freedom to act. A ball in and of itself has no freedom, only the degrees of freedom *a priori* forced on it by the environment. It is only a passive respondent to the environment. From the perspective of *life-here-and-now* the tardigrade seems to challenge this idea. This organism has the ability to turn from life to quasi-death and *vice versa* by responding to environmental cues. It is not a passive object that ends its life when the environment becomes too stressful. It is an organism that

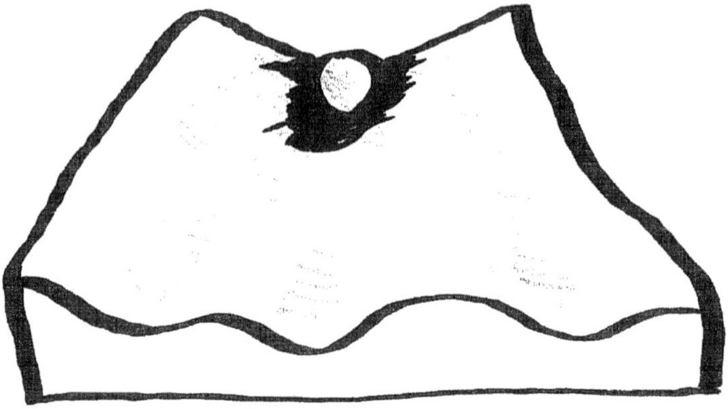

Fig. 15.3 Waddington's landscape.

actively and, if I may add, creatively responds to the stressful environment by turning into a unique state from which it can actively bootstrap itself. If we are looking for a graphical image of a bootstrapping process then we may use the image of the Baron von Münchausen pulling himself out of the swamp by grabbing his own hair (Fig. 15.4).

Indeed, the tardigrade is like the Baron von Münchausen. However, the Baron's image portrays a bootstrapping process as an illogical act. Is there another graphical representation that can do justice to bootstrapping? The visual representation that is perfectly fit to describe the bootstrapping activity of the tardigrade is the *Klein bottle*, a higher-dimensional topological version of the Möbius strip. As I previously explained, the Möbius strip is a one-sided surface in the sense that a bug can traverse the entire surface without crossing an edge.

To review, the Möbius band is interesting because it is non-orientable. In geometry and topology, a surface is called non-orientable, if a figure such as the letter R can be moved about on the surface so that it becomes mirror-reversed. Otherwise, the surface is said to be orientable. A non-orientable surface may allow us to restore the symmetry of an object sliding on it. It is an example of a topology that may allow symmetry restoration and, therefore, reverse computation. Another example of a non-orientable surface is the Klein bottle, which is, roughly speaking, the product of two Möbius strips glued together along each of their lone edges.

What is important to notice about the Klein bottle is that it is a topological structure that passes through itself so that outside and inside meet. For us, it is only important to realize that the idea of bootstrapping is not as illogical as illustrated in the Baron's picture. The movement from the inside to the outside (and *vice versa*), or out of the system into the system, the bootstrapping

Fig. 15.4 The Baron von Münchausen bootstrapping himself.

process, can be smoothly conducted without encountering a paradoxical point of discontinuity. The Klein bottle is the ultimate visual representation of a recursive-hierarchical structure (Rosen, 2004). Moreover, while moving along this re-entering structure we may restore the symmetry of the object we transform. Symmetry is reversible and therefore the Klein bottle is an illustration of a topological structure which is capable of re-entering (and therefore bootstrapping) and for the operation of reversible computation. Can it be that the metabolic network of the tardigrade is built along the lines of the Klein bottle? What does it mean for understanding reversible processes of computation? These questions may sound like a wild speculation but one should be attentive to the relation between thermodynamics and topology, a relation that has been almost exclusively studied by theoretical physicists, and seek for creative ways of answering them. In this sense, rather than providing conclusive answers to the mystery of cryptobiosis, this chapter has the modest purpose of pointing at possible directions for inquiry. It is like a traffic sign pointing in the right direction rather than being the direction itself. As such it should be judged.

In this chapter I tried to deepen our understanding of a recursive-hierarchical structure by using a concrete example. The idea of a recursive-hierarchy is relevant for many other cases and from a reflective perspective even for the writing process in which I am now involved. To quote Deleuze (1994):

> We write only at the frontiers of our knowledge, at the border which separates our knowledge from our ignorance and transforms the one into the other. (p. xxi)

Sliding along the multidimensional topology of our life we cannot but agree with this statement, which points at the writing process as a re-entering process in which our knowledge and ignorance are continuously negotiated.

Chapter 16

Context and Memory: A Lesson from Funes the Memorious

> The very basis of our conscious existence is memory, that is to say, the prolongation of the past into the present, or, in a word, duration, acting and irreversible. (Bergson, 1911, p. 17)

Summary

It is common to think about the adaptive immune system as having a memory. However, memory is always accompanied by the complementary process of oblivion. Is there immune oblivion? In this chapter, I address this question from a meaning-making perspective and suggest that memorization and oblivion are two necessary and complementary processes for meaning making and for attuning us for the context of the here and now. I inquire the implications on this idea for understanding immune memory and immune deficiency among the elderly. This case will help us to better understand the meaning of context.

1. Introduction: "Languaging" in Context

The immune system has been discussed on a theoretical level from a variety of perspectives and with various metaphors (Tauber, 1996, 2002). One possible perspective is the biosemiotics (Barbieri, 2002; Hoffemeyer, 1996; Markos, 2002; Neuman, 2004b; Sercarz *et al.*, 1988). This perspective suggests that we can gain insights into the behavior of the immune system by approaching it as a meaning-making system (Neuman, 2004b), which is continuously involved in making sense out of a variety of signals. Far from being anthropomorphic, the biosemiotics perspective may offer us new ways of examining major issues in theoretical immunology.

The aim of the present chapter is to deepen our understanding of context by offering a new perspective on immune memory. To present this perspective with its full complexity and meaning, I use the linguistic metaphor, introduce the idea of *languaging*, and weave a theoretical thread among language, context, memory, and oblivion.

In one of his essays, the linguistic anthropologist Anton Becker (2000) discusses an insightful observation made by the Spanish philologist and philosopher José Ortega y Gasset. Ortega y Gasset noticed that in natural language we have a delicate balance between manifestation and silence and that "each people leaves some things unsaid *in order* to be able to say others" (Becker, 2000, p. 6). This idea previously presented indicates that the non-present, the hole in the bagel, is no less important than the present, that is, the bagel itself. However, our inclination toward objects usually misleads us into underestimating the importance of the non-present, including silence in language and biology.

Becker examines this observation for better understanding translation between languages. He argues that if silence is an important aspect of language, or more accurately of the language activity that Becker, following Maturana and Varela (1992), calls *languaging*, then we face a problem when translating. Not only do we have to translate what is said; we also have to translate what is unsaid! How can we translate the "unsaid?" How can we translate a silence? To review, Becker's answer is that a translation necessarily misses some things. In this sense, a complete translation is impossible. The difficulty of silence led Ortega y Gasset to suggest that a "theory of saying, of languages", would also have to be a theory of the particular silences observed by different people" (quoted in Becker, 2000, p. 285). This insightful observation draws our attention to the complementarity of speech and silence, the present and the non-present, memory and oblivion. This complementarity is highly relevant to the notion of memory and oblivion in biological systems, specifically the immune system. After all, what is oblivion if not silence? At this point, armed with the idea of memory and oblivion as complementary processes, I would like to turn to another aspect of languaging.

Languaging plays on the strings woven between a person and a context (Becker, 2000). It is an activity of "shaping old texts into new context" (Becker, 2000, p. 9). In other words, languaging/semiosis is the activity in which the abstract and general schemes of memory (*i.e.* the old texts) are sewn into the particularities of context—the "here and now" (to use psychoanalytic jargon).

Here, we come to the point where context (the "here and now") and memory (the "there and then") meet. Languaging involves enacting the past and attaching it to the concrete present. In this sense, languaging is not denotational but *orientational* (Becker, 2000). It is "one means by which we continually attune ourselves to context" (Becker, 2000, p. 288). The same logic applies to sign-mediated processes in biological systems. Signs in biological systems are not denotational but orientational too. For example, an antigen is not a pointer that establishes a correspondence with the non-self (Cohen, 2000a). It is a trigger that in the appropriate context is interpreted in a way that re-orients the behavior of the immune system (Cohen, 2000a; Neuman, 2004b).

From a wider perspective, any semiotic activity, such as human languaging, may be considered a way in which an organism orients itself toward context, the particularities of the here and now. Therefore, context is not only the background of constraints that shape the behavior of the organism but also the particular circumstances toward which the organism attunes itself through semiotic mediation (*i.e.* languaging). It is an activity in which the past is woven into the present—the here and now.

Bringing the past into the present through the semiotic mediation of languaging is crucial for making sense out of texts. Without memorization of texts, we would have to count on grammar and dictionaries (Becker, 2000, p. 287), which are insufficient for meaning making.

The relative impoverishment of dictionaries and grammar in terms of understanding meaning making is also discussed in one of Borges's essays (Borges, 2000b). In "Ezra Pound as Translator: Between Matter and Form", Borges points out that in the Middle Ages dictionaries did not exist and the translator recreated the source text in his own way. This statement is made in the context of translating poetry. Borges says: "Those who, like us, are devoted, with greater or lesser success, to the practice of poetry know that the essence of verse lies in its intonation, not in its abstracted meaning". He then criticizes Pound's critics by saying that "they refuse to acknowledge that his translations reflected not the matter of the original but its elusive *forms*" (Borges, 2000b, p. 51).

Matter is what the text or organism is made of, its basic units, but the form is the elusive and dynamic organization that infuses the matter with life. The meaning of a verse cannot be found by looking in dictionaries or by analyzing the grammar of the verse, just as the meaning of life cannot be found in matter but in the organization that "becomes cause in the matter" (Strohman, 2000). In sum, languaging, embedded in memorization, texts, and communication, is necessary for understanding the attunement to a particular and temporal context. But what is memory? The retrieval of previously known facts? Prolongation of the past into the present?

I prefer to use the term *memorization* rather than *memory*. Memorization can be approached from the perspective of dynamic systems as *the process that within a given subject constitutes a link between the products of* reversible *and* irreversible *processes of computation.*

To review, an irreversible process is one in which the input cannot be restored from the output. A reversible process is one in which the input can be restored from the output (Landauer and Bennett, 1985). If we adopt the classic etymology of *computation—computare*, where *com* means "together" and *putare* means "to contemplate" or "to consider" (von Foerster and Poerksen, 2002)—then the output is the whole emerging from the interaction of micro-level components. A reversible process is a process in which the micro-level components can be restored from the emerging whole. An irreversible process is one in which this backward restoration is impossible. Therefore, an irreversible process involves oblivion and a reversible process involves memorization.

The delicate balance between reversible and irreversible processes is crucial for maintaining the organism (Neuman, 2006). For example, an interesting question concerns the way in which a reversible activation of a cell-signaling pathway leads to virtually irreversible changes in cell fate (Xiong and Ferrell, 2003). It should be recalled that biochemical reactions involved in cell signaling are reversible. How does a differentiated cell that makes an irreversible commitment to a given fate (*i.e.* a muscle or a nerve cell), "remember" its commitment long after the sign/hormone has disappeared? At least in the specific case of a differentiating cell it was found that *positive-feedback loops* play a crucial role in perpetuating the original, reversible signal (Sible, 2003). In other words, the reversible signal turns out to be an irreversible trace through feedback loops. That is, the prolongation of the past into the future is established through feedback loops, and reversible and irreversible processes are linked together.

The idea that memory is maintained through feedback loops suggests that memory is not an activity of *retrieval* but of *perpetuation.* When we ask ourselves how a leaf "remembers" its evolutionary history, the answer is that it does so through the dynamics of feedback loops that constitute its fractal structure. In this sense, memorization is also evident in a snowflake. However, only in living systems memorization is attuned by sign-mediated activity (*i.e.* languaging) to the particularities of context. In living systems, memorization not only echoes the past but also engages in a dialogue with the present. A "dialogic feedback loop" is one that changes its parameters (or its own structure!) in response to the particularities of the here and now (*i.e.* context). Feedback loops that change their own structure/parameters in response to context through semiotic mediation are at the heart of memorization and oblivion. This point will be illustrated with regard to immune memory.

2. Immune Memory

Immune memory concerns the ability of the immune system to remember past pathogens, which is evident from the prompt response of the immune system to the reappearance of past pathogens. More specifically, following exposure to an antigen, naïve T cells undergo clonal expansion followed by clearance of the antigen. This is followed by a phase of contraction, during which virus-specific T cells undergo apoptosis, and then a few virus-specific T cells stabilize and remain as memory T cells. Due to these memory T cells, repeated exposure of the immune system to the potential pathogen will lead to a response that is more rapid and of greater magnitude than the response following the initial exposure. This immunological memory provides the rational basis for protection by vaccination.

Although immune memory is an established behavior of the immune system and is the rationale underlying vaccinations, "the immune cells and molecular machinery responsible for maintaining immunological memory have remained surprisingly elusive" (Mackay and von Andrian, 2001, p. 2323).

However, it has been found that some T cells become memory cells that differentiate into effector cells when they re-encounter an antigen. The memory T cells exhibit differential expression of adhesion molecules and chemokine receptors that allow them to home in on lymph nodes, nonlymphoid tissue, and mucosal sites, and to respond to microbes at peripheral tissue sites (Gupta *et al.*, 2005). Although T cells probably do not enter nonlymphoid tissues, the effector and memory T cells can migrate to nonlymphoid tissues such as the skin or mucosal membranes, where the pathogens are first encountered.

According to this description, immune memory is a *structural* event because the T cells transform into memory cells by acquiring new surface molecules that allow interaction with other cells and the expression of surface proteins that facilitate movement. Although structural changes are the easiest to detect and although the molecular reality is characterized and studied through structural characteristics, it is still an open question whether memorization as a dynamic process is best explained by a structural explanation.

A dynamic explanation would consider immune memory in terms of feedback loops that regulate the apoptosis of clones and the generation of immune cells. The idea that immune memory is maintained by feedback loops is not new; it can be traced back to Jerne's network theory of the immune system (Jerne, 1974; Perelson, 1989).

The idea of feedback loops maintaining immune memory is appealing but problematic in several respects. The main problem is that the network theory of the immune system considers the system to be striving to achieve a given equilibrium. This is a constitutive principle of Jerne's network theory, and its implications for self and non-self discrimination have been discussed by Tauber (2002) and Neuman (2005).

The basic assumption that the immune system strives for equilibrium is not beyond criticism. An equilibrium or homeostasis assumes a *set point*, which is the goal toward which the system strives (Cohen, 2001). For example, body temperature is regulated by a set point of 37°C. When the temperature deviates from the *confidence interval* (to metaphorically borrow the statistical term), irreversible damage is caused to the body.

Cohen (2001) addresses the question of whether the immune system uses a set point, namely, the elimination of pathogens. Cohen's answer is clear: no immune set point exists. In contrast to the popular conception of the immune system as a defense system, he argues that the immune system is actually a *maintenance* system that carries out a variety of daily tasks such as the healing of wounds, tissue regeneration, and waste removal.

Immune maintenance is carried out by *inflammation*—the dynamic processes set in motion by injury that lead to healing (Cohen, 2001). In a case in which the goal of the immune system is defensive, the set point is the elimination of pathogens. However, as is well known to anyone experienced in maintenance work such as gardening, maintenance is an ongoing, continuous activity with no clear set point.

Cohen argues further that not only does no set point exist for the immune system but also that it would have been a catastrophe if one did exist for reactive systems such as the immune system and the brain. Our brain is always reacting, responding, anticipating, and elaborating. Only a dead person's brain has a set point, a fixed attractor. The fact that the immune system has no set point does not mean that feedback loops do not exist but that their meaning is different:

> Set-point systems use feedback exclusively to navigate to their set points; reactive systems [like the immune system and the brain] do not have set-point "goals" to reach. Reactive systems use feedback to adapt their internal images [their habits], to change their patterns of response in accordance with the patterns of signals emanating from their world of interest [*i.e.* context]. (Cohen, 2001, p, 13)

Cohen concludes: "Reactive systems don't aim for set points; they aim for *dialogue*" (Cohen, 2001; emphasis mine).

Cohen actually argues that the immune system is a reactive system that uses feedback loops to adjust to context. It engages in a dialogue with the environment by adapting continually to the challenges of the here and now. This is exactly what I previously suggested. Feedback loops are the mechanism used by the system to bring forth past experience in order to deal with present challenges of the here and now. The immune memory is an instance of this general logic. In this sense, adjustment is the orchestration of reversible and irreversible processes through feedback loops.

3. A Lesson from Funes the Memorious

Returning to the subject of memory, a question that should bother us in the context of immune memory is the appropriateness of using the term *memory* to describe a certain behavior of the immune response. For example, memory, at least in the context of an organism, necessarily assumes *oblivion.* As I previously argued, memory and oblivion are two necessary and complementary sides of the same coin. No cognitive system with limited resources can allow itself to remember everything; when it fails to forget a price is paid. An example of this price can be found in Borges's famous story "Funes the Memorious" (Borges, 1962).

Funes is what we would today call an *idiot savant*, a person who is mentally deficient in general but who displays remarkable aptitude in some limited field (usually involving memory). Funes remembers everything in such detail that the present fades away in favor of past memories. He first acquired his memory as a result of an accident: "On falling from the horse, he lost consciousness; when he recovered it, the present was almost intolerable it was so rich and bright" (Borges, 1962, p. 112).

Funes has a prodigious memory, but he pays a price for his inability to forget: "he was not very capable of thought. To think is *to forget a difference, to generalize, to abstract*. In the overly replete world of Funes there were nothing but details, almost contiguous details" (Borges, 1962, p. 115; emphasis mine).

As Borges insightfully illustrates in his story, the ability to think is associated with a guided oblivion: the loss of information (*i.e.* difference) in favor of higher-level differentiation (*i.e.* a difference that makes a difference). Generalization characterizes not only the human mind but the immune system as well. Without the ability to forget, generalization cannot take place. In living systems, reversibility (memorization) and irreversibility (oblivion) are two necessary processes that must be carefully orchestrated.

The importance of oblivion is illustrated by the effect of aging on the immune system. It is known that the incidence and severity of infectious diseases increase in elderly people even though, paradoxically, they have more clones of memory cells (Akbar *et al.*, 2004). This finding may be associated with less successful immune activity. "One striking feature of the immune system in the elderly is the number of large clonal populations with highly differentiated phenotypes, indicating that cells approaching immunosenescence [age-dependent decline in immune function] might accumulate rather than disappear" (Akbar *et al.*, 2004, p. 741).

The accumulation rather than disappearance of these cells may indicate a failure of oblivion. Does the immune system of elderly people suffer from the same problem as Funes the Memorious?

I would like to suggest that it does. The immune system is a cognitive system that elaborates information and makes decisions in a given context (Cohen, 2000a). It is therefore susceptible to the same cognitive constraints as the human cognitive system.

Is there immunological oblivion? If so, what logic guides it? The memory and oblivion of immune cells can be approached from the perspective of the survival and death of memory cells.

Sprent and Tough (2001) argue that "the turnover and survival of memory cells are controlled by cytokines" (p. 246). This suggestion reminds us of the importance of feedback loops in maintaining memory. At the end of the immune response, the T cells lose contact with the antigen and most of them die. However, a minority of cells survive without direct contact with the antigen. The antigen is a stimulus that is turned into a memory trace by biological feedback loops. In other words, immune memory cannot be fully described by means of simple structural characterizations of the T cells; we also have to take into consideration the feedback loops that maintain this memory through *cytokines*.

Cytokines have essential roles in immunity, including immune cell development, immunoregulation, and immune effector function. Cytokines are described as having "complex actions" and as being "pleiotropic, redundant to some degree, and [inducing] the production of other cytokines" (O'Shea *et al.*, 2001, p. 38). In other words, *cytokines are biological signs*. The semiotic nature of cytokines becomes clear if we discuss them in the context of autoimmunity. According to O'Shea *et al.* (2001), "The same cytokine can promote immune and inflammatory response in some circumstances and inhibit response in other settings" (p. 43). In other words, cytokines, like linguistic signs, are context-sensitive. In one context they mean one thing and in another context they mean another.

Cytokines mediate immune memory by being responsible for the guided death—apoptosis—of T cells. In this context, it was found that several cytokines such as IL-2, IL-7, and IL-15 are involved in this process (Sprent and Tough, 2001). With regard to IL-2 it was argued that "the non-redundant role of IL-2 *in vivo* is to *constrain lymphoid growth* and maintain peripheral tolerance" by promoting programmed cell death (O'Shea *et al.*, 2001, p. 38). The cytokines, at least IL-2, are portrayed as contextual signs that, like any other context, constrain a certain behavior. In other words, they constrain lymphoid growth and thereby create the oblivion necessary for memorization to take place.

In sum, memorization is the activity by which we "continually attune ourselves to context" (Becker, 2000, p. 288)—that is, to the changing and *ad hoc* circumstances of present challenges—by enacting the past. According to this interpretation, memory is not simply a stable state in the system's knowledge base and maintenance of this state by feedback loops but a constraint imposed on the possible variety of present behaviors by past experience. The feedback loops that maintain immune memory are adjusted and regulated by cytokines. The cytokines are responsible for the guided death of cells, and therefore for the oblivion that paves the way to memorization.

4. Immune Memory and Aging

As time unfolds, the individual's immunological memory broadens. Aging is therefore associated with an increase in the number of memory T cells (Aspinall, 2000). Nevertheless, it is known that the elderly are at greater risk of certain infections than younger individuals. The role of the immune system in this susceptibility has been discussed in the literature. It has been argued, for example, that the thymus, which is responsible for the production of T cells, atrophies and its output drops considerably (Aspinall, 2000). Evidence suggests, however, that aging and immunity should be discussed in the context of the delicate balance between innate and adaptive immunity (DeVeale *et al.*, 2004). It has also been argued that aging is associated with a decline in adaptive immunity and an increase in innate immunity. The increase in innate immunity results in chronic inflammation, which may be interpreted as an impaired ability to clear up foreign antigens entirely (DeVeale *et al.*, 2004).

It should be kept in mind that adaptive immunity relies on three types of lymphocytes: B cells, cytotoxic T cells (which identify and mediate the killing of infected host cells), and helper T cells (which rely on antigen-presenting cells and assist the B cells and the cytotoxic T cells). Aging was found to be associated with a reduced proportion of naïve T cells relative to their memory counterparts. It is hypothesized that the alteration in naïve and CD8+ memory T cells in aging may be due *to their differential sensitivity to apoptosis* (Vasto *et al.*, 2006). Some memories simply die hard, and when they do not die they fill up the entire "immunological space" (Franceschi *et al.*, 2000, p. 1719) and prevent the proliferation of effective T cells. From an evolutionary perspective, the integration of findings on immunity and aging suggests that in aging the sophisticated, more recent mechanism deteriorates while the more primitive system is preserved and even boosted. Strong inflammation, which has an evolutionary merit at a young age, becomes, as DeVeale *et al.* (2004) put it, "the enemy within".

Explanations that are based on the dichotomy between the innate and adaptive immune system are deficient. The two systems are embodied in a way that makes their analysis as two separate systems irrelevant.

My explanation of immune deficiency among the elderly is more general and concerns the meaning of adaptivity in terms of reversible and irreversible processes. *Adaptivity* means the ability to be attuned to the context of the "here and now" through semiotic mediation (*i.e.* languaging). Enacting the past through orchestrated processes of memorization and oblivion are one aspect of this adaptivity. Immunosenescence is evident when the immune system fails to "dialogue", to use Cohen's term, or when the sign-mediated activity of the immune system fails to orchestrate the delicate balance between memory and oblivion. The practical implication of this suggestion is that in order to take care of the elderly we should teach the immune system to forget! This counterintuitive suggestion is currently beyond the reach of our interventions, but it may be an interesting hypothesis to examine.

5. Conclusion

Part of the mystery of immune memory for immunologists is that they implicitly rely on their misguided, naïve understanding of the immune memory as some form of computer memory or human memory. Although, metaphors are inevitable in scientific inquiry some of them are simply misleading. Immune memory is not analogue to computer memory or to human memory. These metaphors are misleading and leave questions of biological memory in general and immune memory in particular unsolved. In this chapter, I offered a different conceptualization of memorization in terms of a sign-mediated orchestration of reversible and irreversible processes through feedback loops. This conceptualization is grounded not only in dynamic conceptions of immune memory and the idea of the immune system as a meaning-making system (Cohen, 2006; Neuman, 2004b), but also in our up-to-date understanding of the immune system. In this sense, this chapter, like the previous chapters, is no more than an invitation for examining old problems from a different perspective.

Chapter 17

Transgradience: A Lesson from Bakhtin

Summary

Meaning making involves the ability to perceive beyond the particular and limited perspective of an observer. In this chapter, I discuss this ability—transgradience—from the perspectives of symmetry restoration and dimensionality reduction. We will find again that we are all unique but never alone and that semiosis is what allows us to read in between the lines of the book of life.

1. Introduction: There is No Alibi in Existence

The gap between the abstract and the concrete is at the center of Volosinov's theory of meaning but also at the heart of Bakhtin's manuscript, "Toward a Philosophy of the Act" (Bakhtin, 1999). The gap is framed by Bakhtin in the context of the world and the representation of the world. While the utterance actually takes place in a concrete event of communication, the sentence is a way of representing this event. In this context, and as continuously discussed in the book,

> something is *always left out* of account when we describe our actions. Bakhtin argues this is not merely a weakness in our own powers of description, but a disunity into the nature of things. (Holquist, 1999, pp. x–xi; emphasis mine)

In other words, when we conduct the semantic shift, when we move from the experience to the description of the experience, when we measure and turn information into meaning, something is necessarily lost. This is an inevitable result of meaning making, the expression of an irreversible process of computation, and as will be later discussed of dimensionality reduction.

The question is:

> How, then are the two orders—experience and representation of experience—to be put together? ... how can concepts that by definition must be transcendental (in the sense of being independent

> of any particular experience of they are to organize experience in general) relate to my subjective experience in all its uniqueness? (Holquist, 1999, p. xi)

In other words, the question is how to bridge the semantic gap. Bakhtin addresses these questions from a unique perspective. He is not a naïve realist who assumes an objective world in which the contemplating mind has only an observational role. On the other hand, he is not a post-modernist caricature that conceives the world as a discursive invention of our minds. Bakhtin's position is summarized under the title "there is no alibi in existence" meaning that our representation of the world (whether the inner, the social, the imaginary, and so on) has direct consequences (moral, social, practical, etc.) for being-in-the-world. In other words, the

> active experience of experiencing, the active thinking of a thought, means not being absolutely indifferent to it. (Bakhtin, 1999, p. 34)

A *participatory* representation, to draw on one of Bakhtin's key words, is not a representation that is indifferent to the reality it represents. It is not a representation that tries to hide the fact that it is a representation, but a representation that has no alibi and ought to align the contemplating mind with the actual experience that evolves in time and in concrete contexts of interaction with concrete consequences. The general lesson of this suggestion is that whenever we represent or study an act of communication, we should explain it by using a representation that in some way or another is not indifferent to the particularities of lived experience. That is a representation or a theoretization that takes into account the particularities of the interaction and its value for the communicating agents whether human beings or cells. We will elaborate this point later. Meanwhile it is important to realize that representing is always a process of semiosis hence a process of meaning making.

This idea resonates with Volosinov's theory of meaning. To review, according to Volosinov's definition, meaning is a functional term. It is functional in the causal sense. Meaning is the result of interaction and it is the result of concerted efforts semiotically mediated. There is no meaning without a mediated interaction and there is no point of speaking about meaning without pointing at the result of this interaction. Here we are getting to Volosinov's final point, which concerns the evaluative aspect of the utterance. Volosinov suggests that every utterance is above all an *evaluative orientation*. An utterance is not a communicative picture of the world. Language is orientational rather than denotational (Becker, 2000). It is a way in which we attune ourselves to context, to the particularities of here-and-now, through semiotic mediation. The idea that meaning is orientational emphasizes the active nature of meaning making. Meaning is always actively implied rather than passively given. However, if we realize the singularity of meaning then difficulties are expected.

2. Singularity in Language

As we previously noticed, the uniqueness of the utterance results from the polysemy of the signs composing it and from its contextual nature. The focus on the singularity of the communicative act and the attempt to theorize it has an in-built problem that removed the desperate Saussure from studying the parole to the study of language as an abstract system of signs. This desperate withdrawal is easy to understand. One must be aware to the paradoxical nature of trying to say something general about uniqueness. This paradox is also evident in the work of Bakhtin as we previously reviewed. As argued by Holquist (1990b),

> it is precisely the radical *specificity of individual humans* that he [Bakhtin] is after: a major paradox in all Bakhtin's work is that he continually seeks to generalize about uniqueness. (p. xx)

The paradox is simple: How can we theorize or generalize about uniqueness, about an event that is non-repeatable? I believe that as modern thinkers we should not be threatened by the attempt to discuss the generalization of uniqueness as long as we avoid falling into hasty generalizations that cannot grasp the uniqueness of the event. I would like to address the challenge of theorizing about uniqueness by introducing several new concepts into the discussion. These concepts aim to deepen our analysis and to produce a theoretization which is not indifferent to its corresponding experience.

The first term I would like to introduce is *singularity*. In mathematics, a singularity is in general a point at which a given mathematical object is not defined or a point of an exceptional set where it fails to be "well-behaved" in some particular way (Wikipedia). For example, a point where the function is undifferentiated or the superposition of a particle. My thesis concerns the points of semiosis—signs—and therefore singularity will be used in the sense of a sign at which a given semiotic object is not defined. Can we identify a situation in which a sign is in singularity? After reading the chapter that concerns the polysemy of the sign this question is easy to answer. Before a sign is interpreted in context it exists in a superposition. This superposition is a situation of singularity in which the function of the sign, the concrete signified to which it refers, is not defined. That is, singularity is the state of a sign as a de-contextual unit of language. When "I love you" is uttered, the signs "I", "love", and "you" are singularities. Their meaning is determined only by implying their concrete content in a given context.

Why should we care about singularities? To answer this question let me introduce another concept, which is *projection*. The reason I am presenting this concept is because singularities can result from projections and because as I will argue in the concluding chapter, signs in biological systems result from projections.

Concerning geometrical spaces, *projection* is one way, very obvious in visual terms when 3-D objects are projected into two dimensions and result in singularity. Let me discuss this issue by introducing the term *dimension* (Wikipedia).

In common usage, a dimension (Latin, "measured out") is a parameter or measurement required to define the characteristics of an object—length, width, and height, or size *and* shape. In mathematics, dimensions are the parameters required to describe the position and relevant characteristics of any object within a conceptual space—where the *dimensions* of a space are the total number of different parameters used for all possible objects considered in the model. Generalizations of the concept are possible and different fields of study will define their spaces by their own relevant dimensions, and use these spaces as frameworks upon which all other studies (in that area) are based (Wikipedia). For example, a semiotic system may be described as having dimensions specifying the parameters or the measurement required to define the meaning of a sign. In natural language processing (NLP), for example, the meaning of a word may be defined by the frequency of other words that co-occur with it in sentences. Each word is a vector in *n*-dimensional space and the dimensions are the words that co-occur with it in a sentence. The vector's length is determined by the frequency of co-occurrence.

To illustrate singularity through projection let us use Plato's famous allegory of the cave. In the allegory, Plato likens people untutored in the theory of forms to prisoners chained in a cave, unable to turn their heads. All they can see is the wall of the cave. Behind them burns a fire. Between the fire and the prisoners there is a parapet, along which puppeteers can walk. The puppeteers, who are behind the prisoners, hold up puppets that cast shadows on the wall of the cave. The prisoners are unable to see these puppets, the real objects that pass behind them. What the prisoners see and hear are shadows and echoes cast by objects that they do not see. Figure 17.1 is my own illustration/interpretation of Plato's Cave.

The prisoners can see only the projection of 3-D objects on a 2-D surface. By projecting the bodies to the lower dimension, information in the sense of differentiation is necessarily lost. Points that were separated in the 3-D space are now forced to condense into a single point. *Reduction in dimensionality entails the loss of information*, a loss which is probably irreversible.

Fig. 17.1 Plato's cave.

The prisoners cannot restore the puppets from their 2-D shadows on the wall. *This irreversibility is evident whenever the output of a certain computation is produced on a lower dimensionality from the one in which the input was used without allowing redundancy to be produced.* This idea can be extended to other spaces rather than the 3-D space. To explain this point I will introduce the term *informational landscape* (Cohen, 2006).

The term landscape is usually used in the sense of visual scenery. However, Cohen (2006) uses the term *informational landscape* to denote "an array of information that, like a natural landscape, invites exploration". In such an informational landscape a projection to lower dimension entails irreversible singularity. Let us consider a semiotic network, a semiotic matrix, as our informational landscape. The dimensionality of the matrix is extremely high because of the high degree of *connectivity patterns* that potentially define each sign. Singularity is created when the rich patterns of our mind are projected into the lower dimensionality of communicated language that has to use grammar, polysemy, and dictionaries as platforms. When I curl my little daughter's hair and say to her "Tamar, you are sweet", no grammar or lexicon will grasp the rich, the affectionate, and personal sense from which this utterance was produced. My unique perspective will be necessarily lost when translated into the generality of sign. We may be extremely proud of our language but we should realize that language as grounded in the abstract can never grasp the richness and the uniqueness of our particular experience. The interconnected web that emerges from my own life experience as a father cannot be easily restored from the word *sweet*. Like Plato's cave, only shadows remain. This conclusion should not surprise us. Our thoughts are mediated by signs but signs as communicated between people have to be polysemous and as such indeterminate when uttered as discussed in the chapter concerning the polysemy of the sign. In sum, the need to communicate through signs forces us to use units that in themselves are meaningless. These indeterminate semiotic points are produced when we project the rich patterns of our mind into a lower-dimensional semiotic landscape, the landscape shared by the interlocutors in the situation. How can the information lost in the inevitable process of communication be even partially restored by my interlocutor?

Singularity means the loss of information. However, dimensionality reduction can be compensated for by repetition, which is actually redundancy. This point can be graphically illustrated. When a structure of *n*-dimension is represented by a structure in fewer dimensions then a *repetition* of one or more points occurs. For example, the triangle in Fig. 17.2 is represented in two dimensions.

If we want to represent this triangle in a lower dimensionality as a line then the point A must be repeated (Fig. 17.3).

Repetition as redundancy is necessary if we want to avoid the loss of critical information as we reduce dimensionality. Maybe this is the reason

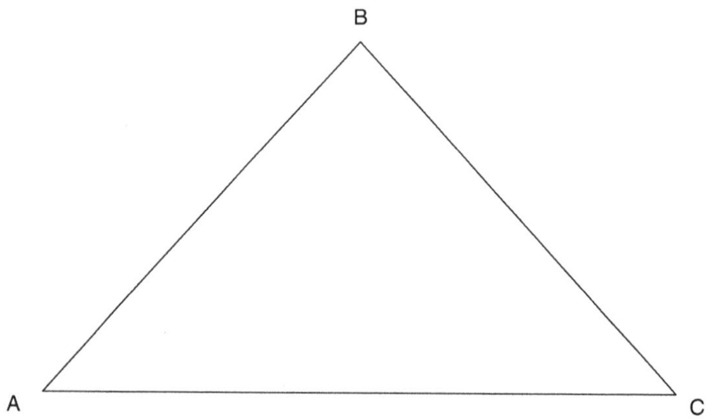

Fig. 17.2 A triangle.

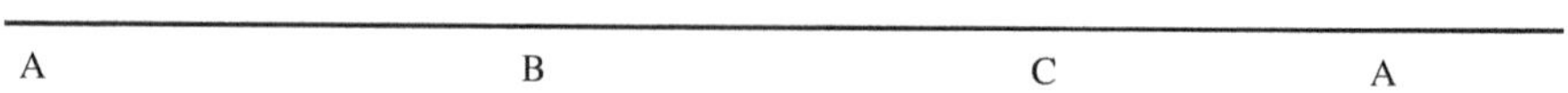

Fig. 17.3 The triangle represented as a line.

why biological systems that are sign-mediated are redundant. In trying to communicate their multidimensional position they necessarily reduce dimensionality and compensate for information loss through repetition. This idea was introduced to psychoanalysis by Ignacio Matte-Blanco (1988) and its relevance for understanding semiotic processes will be discussed later. Now let us move to Bakhtin.

3. Bakhtin on Meaning

The question of how meaning, in the phenomenological sense of a structured form as it appears to the individual, emerges out of fragmented, and even chaotic experiences, has bothered scholars since antiquity.

In his early philosophical writings, Bakhtin (1990) introduces the term *architectonics* to describe how entities relate to each other and the term *aesthetics* to describe how parts are shaped into wholes—for example, in perception, how sensory experience is organized into a gestalt form, or in the social realm, how "I-Thou" relationships are structured.

Architectonics is an activity of meaning making, "making sense out of the world by fixing the flux of its disparate elements into meaningful wholes" (Holquist, 1990b, p. xxiv). However, architectonics is not a structuralist agenda. It is important to notice that Bakhtin was not a structuralist to the extent that he was not seeking to identify de-contextual and static structures underlying phenomena. For Bakhtin the ordered relations between the components of the whole are always in the state of "dynamic tension" (Holquist, 1990b, p. xxiii).

Beings, as Bakhtin never tires of repeating in these essays, is in its essence active: architectonics names the body of techniques by which its sheer flux may be erected into a meaningful event. (Holquist, 1990b, p. xxiv).

This idea of a whole as an organization in a state of a dynamic tension is highly similar to Bateson's idea of a whole and the similarity will be evident in just few lines.

Bakhtin argues further that a whole is always wholeness as long as it is conceived as wholeness by a given observer. In other words, meaning cannot be discussed apart from a contemplating mind and its unique perspective. In a reverse kind of argument we can define the Mind as the system that generates wholes or meaning from a unique perspective. In line with Volosinov we can say that meaning making involves a process of interaction through semiotic mediation in which components/tokens are integrated to produce a synergetic effect we call "whole".

The idea of a whole as observer-dependent emphasizes the notion of meaning making as (a) an active task, (b) a task yet to be accomplished rather than something given to the observer, and (c) the unique and concrete point of view embedded in meaning making. In other words, wholeness, unity of components, structure, or pattern, is not pre-given. Wholeness does not exist in a vacuum independent of a contemplating mind, whether the mind of a human being or the mind of an immune system. Meaning is actively created by mind.

What does it mean that wholeness exists as long as it is conceived by an observer? And what is this observer? Can a molecule be an observer? Can an amoeba be an observer? To address this question let me introduce Bateson's insightful conception of the mind as it was presented in his book *Mind and Nature* (Bateson, 1973). The mind, at least as characterized by Bateson, can be identified with the notion of an *observer*. In this sense, linking Bateson's idea of the mind with Bakhtin ideas of meaning may be a constructive theoretical move.

4. Bateson and the Mind

Bateson presents six criteria of the mind. Let me present these criteria, briefly explain them, and finally link them to Bakhtin's discussion of wholeness.

1. *A mind is an aggregation of interacting parts or components*

The first criterion suggests that mind is not a homogenous entity (*i.e.* spirit) but in our modern terms an emergent phenomenon at one scale of analysis that is created by interactions of components at another and lower scale of analysis. Bateson emphasized the importance of *organization* and *interaction* in explaining mental (*i.e.* non-mechanical) phenomena and he was one of the first to realize, to use Strohman's expression, that "organization becomes the cause in the matter". As Bateson suggested "'Wholes' are constituted by combined interactions". Explaining the emerging mind of a person or a bacterium cannot rely on a description of components just as explaining life cannot rely on the list of genes. This statement seems to be quite trivial in the age of complexity science but I would like to argue that its radical consequences have slipped our mind, or more accurately been repressed from our mind because of the destructive influence they have on our naïve, oversimplistic, cause–effect models of the world. To explain this argument let me turn to Simpson's Paradox.

At a certain scale of analysis the world appears to us as a set of objects with simple causal relations between them. In this world, the behavior of a variable can be predicted from the values of another variable without too much effort expended in understanding the interactions between the explanatory variable and other variables. In this world, a threatening sound may be an excellent correlate for a danger that should be avoided; the existence of vegetation may be an excellent correlate for the existence of a water source, and the exposure of gums and teeth may clearly indicate to a male gorilla that the King Kong of his group does not like him messing with the females. This is, to use the Peircean term a world of *habits*, simple correlations or associations between variables that allow to us and to other creatures to anticipate and to survive. Not too many complexities are assumed. However, the mind that is able to detect these habits cannot be analyzed as the result of simple habits but as emerging from micro-level interactions in which our naïve Newtonian philosophy does not seem to hold. In their search for lawfulness and order, human beings have studied systems that for most of their evolutionary history have been beyond the scope of their awareness. These systems, such as the brain, are constituted through micro-level interactions but these interactions are complex and cannot be studied through simple causal models.

To illustrate the complexity of interaction let me present Simpson's Paradox. Simpson's Paradox suggests that "an association between a pair of variables can consistently be inverted in each subpopulation of a population when the population is partitioned" (*Stanford Encyclopedia of Philosophy*; Entry: "Simpson's Paradox"). Let me give you an example. A medical treatment can be associated with a higher-recovery rate for treated patients compared with the recovery rate for untreated patients. Yet, treated male and female patients can both have lower-recovery rates when compared with untreated male and female patients! This shocking finding actually suggests that if we have a simple causal model of the world in which a variable *A* is the alleged cause of variable *B*, there can always be another variable *C* that when entered into the game will reverse our causal direction. In other words there is always a factor that "screens off" any correlation. It is amazing that people dealing with the complexity of mind, such as some psychologists, usually ignore this devastating conclusion. Try and ask your colleague psychologists how many of them are familiar with the paradox and understand its relevance for their experimental work.

When things get tough in terms of complexity, our old habits cannot ensure a valid inference. When inquiring into emerging wholes from heterogeneous parts our simple causal thinking is simply irrelevant. The behavior of systems like human societies, the immune system, or the human mind cannot be explained without paying close attention to organization and interactions. Reductionist science has no way of explaining these systems.

2. *The interaction between parts of mind is triggered by difference, and difference is a non-substantial phenomenon located in neither space nor time*

Bateson's second criterion concerns his idea of a difference. In contrast with the physical world in which components act through forces on other components, the world of living systems is a world in which a *relationship* between two parts (or between the same part at different times) activates a third component that Bateson described as the *receiver*. This receiver is activated by a difference or a change. In other words, the basic and constituting unit of the mind is not the duality of sign-object but the triad that includes a receiver that responds to the difference between two other components.

Bateson emphasized the idea that a difference "being of the nature of relationship" is not located in time and space like an object. This is the reason why a difference is the building block of any "mental" system. Even more important and relevant to the thesis I am propagating in this book is the relation between difference and meaning:

> We are discussing a world of *meaning*, a world of some of whose details and differences, big and small, in some parts of that world, get *represented* in relations between other parts of that total world. (Bateson, 1979, p. 99)

This statement links Bakhtin's idea of wholeness with Bateson's ideas of mind and meaning. Meaning is equated with the whole that emerges from the representation and integration of differences.

3. *Mental process requires collateral energy*

The energy used by meaning-making systems is used in a different way than in mechanical systems. Bateson (1979) argues that in life there are two energetic systems:

> One is the system that uses its energy to open or close the faucet or gate or delay; the other is the system whose energy "flows through" the faucet when it is open. (p. 102)

To better understand this idea we can think about a meaning-making system as a hierarchical system is which the flow of energy at level *A* is being channeled by a higher level *B*. Level *B* is the regulatory level of level *A*. For example, the random movement of molecules can be used to do some constructive biological work as described by the ratchet model in Chapter 14. This idea brings us to the next criterion.

4. *Mental process requires circular (or more complex) chains of determinations*

The hierarchical nature of a meaning-making system does not imply a simple top-down determination the same as its reliance on micro-level interactions does not imply a simple bottom-up determination. As argued by Bateson (1979):

> The organization of living things depends upon circular and more complex chains of determination. All the fundamental criteria are combined to achieve success in that mode of survival which characterizes life. (p. 103)

Previously I presented and elaborated upon the recursive-hierarchical organization/dynamics that characterizes living systems. This unique form of organization requires collateral energy in the sense that sources of energy that exists at the physical level of analysis are being used to maintain biological "switches" that in their turn recursively control the flow of energy that maintain them. The mind is a recursive-hierarchical structure.

5. *In mental processes the effect of differences are to be regarded as transforms (i.e. coded versions) of the difference which preceded them*

The rules of such transformation must be comparatively stable (*i.e.* more stable than the content) but themselves subject to transformation (Bateson, 1979, p. 110).

Bateson (1979) argues that:

> Any object, event, or difference in the so-called "outside world" can become a source of information provided that it is incorporated into a circuit with an appropriate network of flexible material in which it can produce changes. (p. 110)

In other words, differences are transformed/coded to causes as they are incorporated into a network of circuits in which they produce a change. An emergent whole is none other than an aggregate of information, a pattern.

6. *The description and classification of these processes of transformation disclose a hierarchy of logical types* immanent *in the phenomena (emphasis mine)*

Where does our whole exist? Bateson's answer is clear, the emergent whole exists in between the levels. Mind cannot be reduced to a single level of analysis. This hierarchy is *immanent* to the phenomena in the sense that a reductionist move is impossible. Following this analysis, the nature of the observer is evident. The observer is also a whole that emerges from multi-scale interactions.

5. The First Law of Human Perception

Bateson's theory of the mind allows us to define the terms: *mind* and *observer*. In this section we will learn that one of the unique characteristics of the observer-mind is that it is always perspectival.

The unique perspective of the human observer is depicted in what Holquist describes as "the first law of human perception": "whatever is perceived can be perceived only from a uniquely situated place in the overall structure of possible points of view" (Holquist, 1990b, p. xxiv). That is, every human being (and we may extend this argument to observers in general) has a unique perspective, which we may describe as the "individual" part of his or her consciousness[1]:

> When I contemplate a whole human being who is situated outside and over against me, our concrete, actually experienced horizons *do not coincide*. For at each given moment, regardless of the position and the proximity to me of this other human being whom I am contemplating, *I shall always see and know something that he from his place outside and over against me, cannot see himself*... As we gaze at each other, two different worlds are reflected in the pupils of our eyes. (Bakhtin, 1990, pp. 22–23; emphasis mine)

The two different worlds that are reflected in the pupils of our eyes are the worlds that will necessarily be reduced to lower dimensionality in order to allow for communication. The dimensionality of my own unique place as an individual cannot be replicated. Sheep can be cloned (to a certain extent), paintings can be copied, but my own uniqueness can never be reproduced. Mind is always perspectival.

A whole is always a positional or a relational structure. It is a relational phenomenon as long as it expresses coherence with regard to a given set of coordinates. The singularity of the individual, its unique place in existence, is the vertex of this system of coordinates. Does it imply a cacophony of mutually exclusive perspectives or in contrast homogenous and closed Leibnitzian monads? And if this is the case how do different minds converge into the same perceptions and bridge their ontological autonomy? The answer is given in the next section. Meanwhile, let us remember that Bakhtin

---

[1] In this chapter, I use the term *consciousness* in the sense of a mediating semiotic system.

was highly interested in the scientific developments of his time (to include Einstein's physics) and his solution to the above questions can be reframed within the much broader context of science. As we will see later, Bakhtin was not only echoing the Zeitgeist of the emerging new physics, but has in his own way made a potential contribution to our understanding of the biological realm.

The "general law of uniqueness" stresses the existence of the mind as a unique being that is clearly demarcated from other minds. An important aspect of this law is the singularity of the observer as manifesting its unique coordinates within a phenomenal space. As suggested by Deleuze (1994, p. 25), "A dynamic space must be defined from the point of view of an observer tied to the space, not from external position".

Again the positioning of an observer in a given situation cannot simply be considered to take place in a 3-D physical space. To review, the mind is constituted as a recursive-hierarchy of transformed differences, and differences do not exist in time and space. They are relations. The differences exist in a recursive-hierarchy, meaning that different orders of dimension interact with each other and allow bootstrapping to occur. To review, an imaginary thief that exists in the fourth dimension can steal a diamond from a 3-D safe without breaking the safe (Rucker, 1977). Along the same lines, a mind that exists as a recursive-hierarchy can penetrate itself, observe itself, and bootstrap itself without destabilizing the lower dimensionality on which it operates. It is like the 4-D thief, who can steal the diamond from the 3-D safe without breaking the safe: no dynamite, no explosives, just a quick intrusion from the fourth dimension.

My unique perspective results not only from the unique configuration of my semiotic matrix, but also from the fact that as a contemplating mind, I and only I have the ability to look inside myself. In this sense, the positioning of the individual cannot be discussed in terms of a physical space but in terms of a phenomenal space or what may be described as our personal recursive-hierarchical informational landscape.

An informational landscape is not "objective" scenery. It is not the Greek cosmos. It is a landscape of meaning making for a particular observer, for a particular mind. To quote Holquist (1999):

> The difference between the objective cosmos and our human world was brought home to Roman legionaries every time one of their units was punished with decimation: in the order of numbers, the difference between "nine" and "ten" is purely systemic; for the solider standing ninth in line it meant life, whereas the "objective" fact of being tenth consigned the next man in line to death. The difference between that event as seen from the perspective of

> number theory alone and what it meant to an actual legionary on a particular day is the lack Bakhtin's non-alibi seeks to accommodate. (p. xiii)

Let me illustrate this idea with another Roman example. For the philosophers' God, who is everywhere and therefore lacks a specific perspective, there is no difference between a thumb pointing upward and one pointing downward. They are symmetric and therefore identical. See Fig. 17.4, for my painting of the hand gestures.

However, for the defeated gladiator in the Roman arena, the difference between the two signs was a matter of life or death. If the emperor turned his thumb downward, the gladiator's fate was sealed. In this sense, adopting God's perspective is losing perspective, and this is a crucial point religious (or scientific) fundamentalists miss when they try to speak in the name of God (or in the name of other ultimate perspective such as Evolution). The phenomenal space in which meaning making takes place is a set of relations and transformations within which the mind is patterned and not an abstract space in which it is a neutral observer.

Following this line of reasoning, the unique position of an observer in the situation cannot be underestimated, since it encapsulates his or her unique perspective on the situation (*i.e.* the unique patterning of the situation) and therefore determines the identity of the situation and the meaning of a sign or an utterance communicated within it.

Here we return to the first law of human perception. Observers/minds are *asymmetric* in the sense that what one can see from his or her perspective is not what the other sees. *The asymmetry of perspectives is a basic property of all minds that constitutes their systemic closure through active self-differentiation.* "I am" first of all because I am not the world. As realized by Bateson, Spencer-Brown, and Saussure, difference is at the heart of existence and since

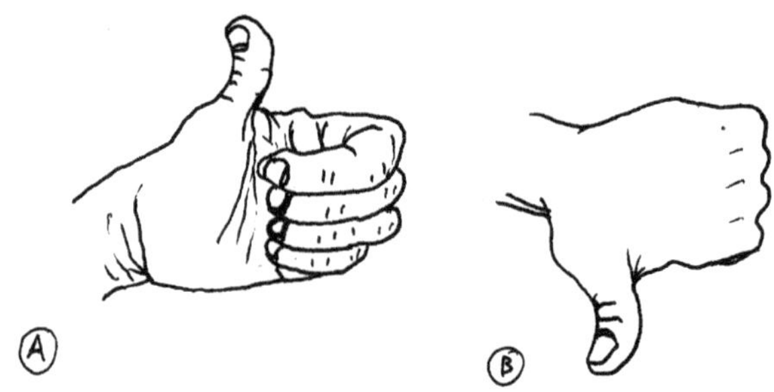

Fig. 17.4 The asymmetry of the hand gestures.

difference is always a relation within a triad, it necessarily generates asymmetry and the first law of perception.

This analysis invites a question. If the asymmetry of observers is built into their unique patterning in the world, how can they communicate their perspectives and coordinate the meaning of a given utterance? After all, in order to understand how a certain utterance is patterned in a given situation and to overcome the asymmetry of their perspectives, interacting agents must restore symmetry of perspectives.

Let us examine this difficulty and its resolution through the symmetry of geometrical forms. A body or spatial configuration satisfies the criterion for symmetry if it "can be superimposed on its mirror image within a given dimensional frame of reference" (Rosen, 1994, p. 17). For example, the two faces in Fig. 17.5 are symmetric since they can overlap through translation along the *x*-axis.

Metaphorically, we may argue that a situation is symmetric if it can be imposed on its mirror image when the positions of its interacting agents/units are interchanged.

In some cases, two objects are asymmetric but their symmetry may be restored through use of a hyperdimension. For example, the faces in Fig. 17.6 are asymmetric, but their symmetry can be restored by means of a rotation in the third dimension.

Unfortunately, for asymmetric objects in three dimensions, there is no hyperdimension through which their symmetry can be restored (Rosen, 1994). The same may be argued metaphorically concerning communication between people. It is commonly held that a person is interwoven in a concrete

Fig. 17.5 Symmetric faces.

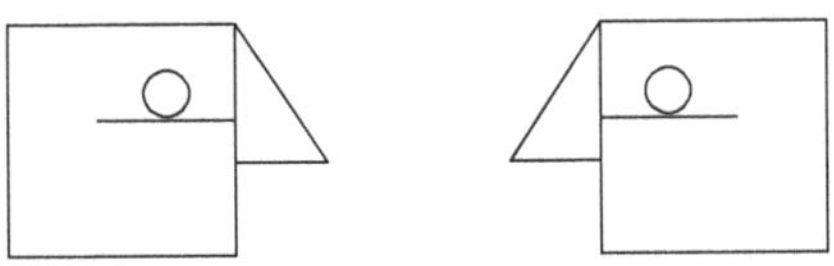

Fig. 17.6 Asymmetric faces.

situation and there is no *hyperdimension* or meta-perspective that can move him or her out of the situation. Therefore, in a case of asymmetry of perspectives, there is no hyperdimension that would allow restoring the symmetry of the communicating semiotic matrixes. This conclusion is wrong. Languaging provides us with a way to transcend our unique positioning. Through language we are able to restore the informational landscapes of other minds. We are able to imagine "what is it like to be a bat". We can dream of better worlds and we can hope for a better future. This ability is not the venture of a single mind. As the next section teaches us, we are all unique but never alone.

6. We are All Unique but Never Alone

In contrast with what may be mistakenly inferred from the uniqueness of the individual, the "general law of uniqueness" does not imply a solipsistic stance according to which the self takes precedence over others, both ontologically and epistemologically. Solipsism is the philosophical stance that one's self is the only thing that can be known with certainty and that *only* one's self exists. Descartes' "cogito ergo sum" is an expression of a solipsistic stance.

Bakhtin is not a solipsist. On the contrary, Bakhtin transcends the solipsistic and dualistic stance that may be inferred from the uniqueness of the individual. He argues that we realize our uniqueness *only* through the existence of others: "We are all unique but never alone" (Holquist, 1990b, p. xxvi). This position is embedded in Bakhtin's opposition against binaries and his conception that wholeness is achieved by *simultaneity* of perception (p. xxiii). Only the simultaneous perception of multiple components from complementary perspectives (as will be further discussed) allows us to see the forest, which is the *whole* of the trees. This is what transgradience is all about. *Simultaneity of perspectives merged together to achieve an integrated and full understanding*. Remember the simultaneity that underlies quantum interference? Simultaneity is crucial for meaning making.

7. From the "I" to the Other

Where does the other enter into this dynamic? Bakhtin begins by assuming that because of the uniqueness of each person, one's perspective is always limited. In other words, uniqueness is a source of both strength and weakness: I can see what you cannot see, but I cannot see what you, as an outsider, can see. Therefore, we need the other in order to obtain a complete picture of ourselves from the *outside*. The idea of the other as a part of my mind necessarily implies, as both a social and an epistemological imperative, a transformation of a human being from his or her unique position to the position of the other:

> I must empathize or *project myself* into this other human being, see his world axiologically[2] from within him as *he sees* this world; I must put myself in his place through the excess of seeing which opens out from this, my own place outside him. I must enframe him, create a consummating environment for him out of this excess of my own seeing, knowing, desiring, and feeling. (Bakhtin, 1990, p. 25; emphasis mine)

How is it possible to see through the other without breaking the boundary between the two differentiated systems? What does it mean to project myself?

8. Signs as a Bridge between the "I" and the Other

The basic question is, of course, how we can transcend our unique point of being, our systemic closure, the boundary of our individuality, and project ourselves onto another person's mind in order to reflect on our unique existence. Bakhtin's answer is: through the power of signs to carry things over from the realm of the individual mind into intersubjective territory, to the collective part of our mind. In this sense, natural language as a sign system is what constitutes the bridge between our individual and collective forms of consciousness, and what constitutes the wholeness of my individuality. In other words, the inter-social semiotic aspect precedes and constitutes my individuality. This suggestion emphasizes the notion of mind as semiotic activity. Since all semiotic/mental activity is necessarily social (Volosinov, 1986), our ability to reflect on our existence is bounded by *ideologies* (*i.e.* systems of signs) shared by the collective. In this sense, our mind reflects not only our individuality and unique position in the world, but also the collective semiotic systems (*e.g.* religious, moral, scientific, and biological) that constitute it (Volosinov, 1986).

Projecting myself to the other by using language is projecting myself into a different mind, a different language, and a different semiotic matrix. As

---

[2] Axiology is a keyword in Bakhtin's thought. Holquist (personal communication) suggests that it refers to the constant dialogue between the given and the created that constitutes our active life as ethical subjects.

suggested by Ortega Y Gasset:

> In reading a distant text, one tries to project oneself not into another mind—at least at first—but into another language, which held grip on that other mind, that other person, who inherited with his language, choicelessly, the greater part of the ideas by which and from which her or she lives, without thinking about them at all. (Quoted in Becker, 2000, p. 371)

9. Conclusion

Mind is not the property of a given individual or the system that uniquely characterizes human beings. Mind is the name we give to a specific type of a system, which is characterized by Bateson's criteria. Mind has a unique point of view, a perspective determined by its unique positioning in a phenomenal space. The ability to transcend its boundaries through languaging is an inbuilt feature of the mind, a feature that involves the ability of self-observation through the others. To a certain extent, human consciousness is a specific expression of this general feature that characterizes different minds, from the mind of the amoeba to the mind of a primate. The use of polysemous signs involves uniqueness resulting from the "empty" nature of the signs. Signs in themselves mean nothing and a concrete content must be poured into them in order to turn the dead string into a live utterance. The singularity of the sign results from the projection of higher-order dimensionality onto a concrete yet to be communicated token—the sign. This projection involves the loss of information and symmetry, which can only be restored by my interlocutor through the projection of himself or herself onto my semiotic matrix. This is a process in which the interlocutors reciprocally and in concerted efforts construct the meaning of an utterance by simultaneously folding layer upon layer of their limited perspective to achieve a global understanding of the situation. This process involves what I previously described as coordination-under-constraints. Singularity is resolved when the higher dimensionality of my semiotic matrix is restored, and the higher dimensionality is restored not when a sign is loaded with a given signified but when the utterance as a whole is woven in between the interlocutors. This lesson should be extended beyond the realm of human communication. A sign can mean nothing for the immune system without the joint effort of immune agents that attempt to transcend their own unique perspective to achieve a global view of the situation. Transgradience is a defining aspect of meaning making. Life is saturated with meaning making and rather than discussing the book of life with its dead letters and signs, we should carefully examine the way in which concerted efforts proceed step-by-step to constitute meaning.

Cat-logue 4

Dr. N: Bamba, my dear cat, did you ever had the feeling that what you experience is beyond words?

Bamba: In my case, beyond yawning, but sure. In fact, it does not surprise me that the gap between our experience and the description of our experience is unbridgeable. Description is possible only through this gap. But remember Bakhtin's lesson: There is no *alibi* in existence.

Dr. N: What does it mean?

Bamba: It means that our representation of the world, which is always sign mediated, has consequences.

Dr. N: Sounds trivial. No?

Bamba: Unless you understand that representing the world is actually meaning making. Now we understand why there is no *alibi* in existence. The way we make sense out of the world has direct consequences for being in the world. If the immune system fails to make sense then death might prevail over life.

Dr. N: Here I'm getting into trouble. Our existence depends on meaning but meaning as taught by Volosinov is always a unique event. So how can we or any other mind survive in an environment of singularities?

Bamba: Paradoxical as it may sound, we can survive only because we use singularities.

Dr. N: What do you mean?

Bamba: A sign is singularity before interaction in context (*i.e.* measurement) determines its value. Singularities result from projections onto lower dimensionality. Think about the sign "I". This sign, which is an enormously rich matrix of thoughts, feelings, etc., is what you use to represent yourself. But when you say "I" how much of this richness, how much of this higher dimensionality is left? The richness of the self as an experience is lost for the abstract and singularity of the "I" that denotes the first person pronoun.

Dr. N: You know, Bakhtinian scholar Michael Holquist wrote something insightful about the "I". He says, "Much as Peter Pan's shadow is sewn to his body, the 'I' is the needle that stitches the abstraction of language to the particularity of the lived experience".

Bamba: Beautiful! Now you understand the need for singularity. We project onto lower dimensionality in order to communicate the particularities of experience through the abstractness of the description.

Dr. N: Has it something to do with Bakhtin's theory of meaning?

Bamba: Definitely! Remember the first law of human perception? This law is actually the first law of the mind. Mind is always perspectival. It is asymmetric. What one can see from his perspective the other cannot.

Dr. N: So how can symmetry be restored? Is there a hyperdimension?

Bamba: Languaging my friend. Languaging is the answer.

Dr. N: Signs again?

Bamba: Indeed. We are all unique but never alone, and we are never alone as long as we can dialogize. We realize ourselves as living systems through others who provide us with the simultaneity of perception and allow us to restore symmetry and to extend the limit line of our minds.

Dr. N: Very simple.

Bamba: But very complex.

PART 4. FROM MECHANICS TO POIESIS

Chapter 18 The Poetry of Living

Cat-logue 5

Chapter 18

The Poetry of Living

What does it mean to understand? This simple and naïve question has no simple answer. Criteria such as prediction, control, or the ability to represent the phenomenon we would like to understand are common answers given to the above naïve question but they are not enough. For example, control is one of modern science's favorite criteria for understanding. If I can control something then I understand it. If you knock out a gene and notice a phenotypic change then you understand the gene. The impoverishment of this conception, or more accurately of this misconception, is an obstacle to understanding life itself. I can control the life of another living creature by simply taking it from him. Taking the life of another creature is the ultimate level of control. Does the ability to control life by taking it indicate that we understand life? Smashing a bothering bug and ending its life is a trivial activity of control but does it indicate that we understand what turns matter into a living and flying bug? Control should not be confused with understanding.

One of the potential obstacles for understanding can be found in the very models that mediate our understanding of the world and the wealth of metaphors we adopt from language activity. Considering the Genome as the book of life is just one of the prominent metaphors we encountered in this manuscript. However, books and the idea of the novel are limited metaphors in understanding living systems. Living systems are not biological novels, they do not have an author, and the question "who is reading the book of life?" cannot be answered by the rather cryptic answer "the organism itself". If we would like to use linguistic metaphors we should use those that are relevant as possible to the fact that our bodies and the bodies of other organisms are matter imbued with life. By using the expression "matter imbued with life" I do not commit myself to any vitalist standpoint. Life is the concept I use to describe the difference between matter and organisms. Although this is a circular argument it cannot be avoided. One may use other concepts instead of life, such as soul or mind (in the Batesonian sense) without causing any harm to the general and commonsensical observation that living forms, although made out of matter, establish a unique category qualitatively different from matter. This point can be illustrated by an amusing anecdote. Several years ago, I participated in a conference on astrobiology and the origin of life. The majority of the speakers were orthodox biologists, biochemists, and physicists. However, one of the keynote speakers was a prominent philosopher who opened his talk with the provocative statement: "The existence of the soul is an undeniable scientific fact". This provocative statement was a trap for the naïve mechanistic thinkers in the audience and as can be expected someone quickly took the bait. A biochemistry professor immediately jumped on his feet trying to control his rage from the fact that into the sacred temple of scientific discussion entered a term that belongs to the despised terminology of religion and poetry. This professor stared at the philosopher with disrespect and asked: "Can you prove it?" The philosopher, who was a student of Sir Karl Popper, gazed at his mechanistic colleague with amusement and replied in a heavy British accent: "My dear Sir cannot you differentiate between a dead and a living person? This is what soul is about." The biochemist, who was probably waiting for a proof in terms of metabolism, sat back on his chair without saying a word.

Understanding life through language obliges us to abandon our naïve instrumental conception of natural language and to turn into a more basic, profound, and general semiotic perspective. There is no author of the book of life and the "language of the genes" does not deliver any message from a "sender" to a "receiver". Human language is just a specific instance of a more general logic of semiosis and as such it should be judged. As an aside, I would like to comment on the functional perspective of language. It seems that the functional conception of language originated from Anglo-Saxon armchair philosophers who used language to command their servants and through this use portrayed language as functional in nature. I can imagine John Austin commanding his servant: "Philip, may I get my glass of brandy" and thinking to himself: "Hmmm, language is a wonderful way to do things. Why shouldn't I write a book about how to do things with words?" Using language to do things is just one aspect of languaging; a totally different perspective will be presented below.

Surprisingly, an illuminating insight into the non-instrumental nature of language can be gained from classical Indian poetry (Shulman, 1986). In the Veda—the most ancient text of classical India—language is not conceived as delivering information, naming phenomena, mimicking the world, or determining meaning. Language is, first of all, not a human product but something that *reveals itself in the human*. We are not simply the masters of our language who, through intentional and rational acts, use it to communicate a specific and well-defined mental content. As we have experienced more than once, in much of our daily use of language we know what we wanted to say only after the conversation has ended, and in many cases we do not really know what we wanted to say even after the conversation ended.

The idea that language "reveals itself" should raise our consciousness to the fact that we as human beings think THROUGH languaging, act THROUGH languaging, and live THROUGH languaging, rather than simply by using language. The capitalistic privatization of life does not ignore mental life or the linguistic realm. We are portrayed as the owners of our thoughts and the masters of our language. The Veda has a different perspective.

The idea that language reveals itself also undermines the role of nouns in languaging. Nouns are not the most important aspect of languaging because in most living systems the correspondence between the sign and the signified is not an issue. As I argued elsewhere (Neuman, 2003a), we live in a reified universe in which nouns have precedence over verbs and objects have precedence over processes. However, modern and Western theories of language and meta-language cannot be used as a valid model for understanding the realm of the living. Nouns do not exist at the cellular level and most of the organisms do not bother themselves with the relation between a given sign and a given signified. Signification in living systems is in most of the cases Vedic style. It is signification in which signs are functional generalities that are the expression of an underlying dynamic. Signs are *semiotic attractors* (Neuman, 2003a) and not the counterpart of mental concepts. This is an important point. The process of biological signification, in a similar way to ancient poetry, is grounded in dynamic *rhythms* and not in static Platonic forms.

Let us learn more about the Vedic perspective. There are four key concepts that define the meaning of language for the Indian poet. The first is that language is not produced but uncovered. It reveals itself to those who can open themselves. This idea leads to the second concept, which is some kind of gentleness or softness that one should adopt in order to understand the underlying reality. Softness should not be mistakenly confused with some kind of weakness. Softness is not only a pre-condition for language to reveal itself but also the basic property of living matter. To review, biological matter is actually composed of *soft matter* that endows it with flexibility and elasticity. Soft matter is different from hard matter such as salt or metal. It is also different from liquids. Being soft does not mean being fluid. Soft matter, like a cell's membrane, involves multi-scale and complex behavior and it has variety of response functions. The Veda and modern physics converge in understanding the importance of softness for understanding life. The multi-scale organization of matter, or the states we call soft matter, is a pre-condition for sign activity to reveal itself. In a universe of hard matter or fluids, life could not have emerged because signs need the pre-condition of softness. As we previously learned, the multi-scale of living systems assumes boundary conditions and signs are those boundary conditions that constitute life. Life reveals itself in soft matter through boundary conditions, through semiosis.

The third constituting feature of language as portrayed by the Veda is the idea of a break. The world was created by a break in unity. An ancient myth that appears in the Veda tells us that Prajapati, the androgynous being and the primordial lord of creatures, felt an enormous emptiness when he gave birth to the world. This horrible emptiness caused Prajapati to re-unite with his offspring by swallowing the newborn. When the newborn saw the empty mouth of his father he shouted in horror. The birth of language is in horror. The horror of separation and our struggle to constitute our autonomous existence: to constitute a self and non-self differentiation. The idea of language as a break (the semantic break?) is expressed in the fact that our first linguistic activity as newborns is to cry. We are born to the world with a cry as our first linguistic expression.

The idea of creation by symmetry breaking is just a modern and scientific form of this ancient metaphysical principle. Recall from Maxwell's demon that a break in symmetry, the emergence of information in the sense of differentiation, is meaningful only for an intelligent "demon" that can create information by memory and comparison (*e.g.* fast and slow molecules). Languaging creates a break between us and the world as we experience it, but this break can lead us, paradoxically, back into the underlying plane from which language emerged.

Languaging is deeply associated with the general order in which we are embedded. This insight is repeatedly enacted by poetry. Poetry has a sacred status in classical Indian culture not because of what it says, which is usually partial and limited, but because of what it does not say—*silence*, and the way it hints through cues to the deep and hidden layer of existence. Personally this is what makes me so excited about reading poetry. Poems by Szymborska and other great poets are a window to a hidden reality and in this sense poetry and science have a common denominator.

In this context the poet is not just an entertainer of the audience, but someone who takes a part in creation and in opening our mind to underlying existence. By imagining through languaging, the poet (like the scientist) enacts a world hidden from our daily conception. We do not enact the world by pointing at it and, as we previously suggested, languaging is not denotational but orientational. It orients us through cues. Meaning cannot be found in any direct correspondence between the language and the world but through the delicate way languaging orients us to context. Here we say no more than has been repeatedly mentioned in this book. As Borges reminds us, to understand a verse we should read in between the lines and listen carefully through the words to the non-present. In this concluding chapter, I would like to discuss the meaning-making perspective offered in this book by poetically reflecting on the complexity and mystery of the semiotic processes that constitute life.

Let us begin with complexity. The world is a complex place to live in. This statement sounds like street wisdom but what does it actually mean? What does it mean when we say that the world is complex? We may address this question with the term "Information landscape" coined by my colleague Irun Cohen. Information landscape is the "maze of information available for potential exploitation by a suitable system" (Cohen, 2006). The complexity of the environment results from the *dimensionality* of this maze. The maze in which we live is a multidimensional matrix of information, and different creatures harvest different niches within this maze. The dimensionality of the maze is a key for understanding its complexity. One of the thinkers who realized this idea was the quantum physicist David Bohm. In a highly similar way to Bateson, Bohm (1998) considered order in terms of *differences of similarities* and *similarities of differences*. In this context, complexity involves ever increasing dimensionality of differences and similarities:

> Now, the simplest curve is a straight line. Here the successive segments differ only in position, and are similar in direction. Then comes the circle; successive segments also differ in direction. But the angels between them are the same, so that the differences are similar. However, the *similarities* defining the circle are *different* from those defining the straight line. This, in fact, is the essential *difference* between the two curves ...
>
> Evidently, it is possible to go on to higher-order differences, whose similarities generate a series of *ordered* curves of ever greater complexity. (Bohm, 1998, pp. 7–8)

The higher the organisms are on the evolutionary scale, the more efficient they are in harvesting this landscape and the more time they invest in learning how to adjust to the landscape. Complex creatures make more complex environments that demand more time to adjust to. The learning process in a modern information-based society is much longer than the learning process in agricultural societies.

The multidimensional nature of the information landscape is a source of opportunity. However, it is also a source of burden because we, and other creatures as well, cannot grasp the totality of the interconnections that constitute the web. Living in a matrix while being aware of it not only puts a sharp limit on our understanding but might, under certain circumstances, result in paranoia. The Wachowski Brothers' trilogy, *The Matrix* is a clear symptom of this post-modern paranoia. Indeed, we live in a matrix in which the ultimate understanding is a fantasy attributed to a super mind like the architect in *The Matrix* or God in the monotheistic religions. For less than the super mind only partial understanding is possible.

Our inability to grasp the totality of *The Matrix* entails a dimensionality reduction as a cognitive must.· The world as represented in mind, whether the mind of a human being or the mind of the immune system, is not a reflection of the world but a *re-presentation* of the world in the constructivist sense. This statement should not be confused with any form of naïve realism. The topology of the mind is similar to the topology of the Klein bottle in which our discrimination between outside and inside collapses. If a bug traverses on the surface of the Klein bottle it will never cross a point of discontinuity that signifies the boundary between inside and outside. So what is the meaning of representing the "outside world" along the lines of the Klein bottle? Let us recall that the Klein bottle is a non-orientable 4-D surface. It can restore the symmetry of an image sliding on its surface. The ability to restore symmetry in a recursive-hierarchical structure is what differentiates between inside and outside. That is, *differentiation results from the dynamics between reversible and irreversible processes.* For living creatures the world neither exists as separated from our mind, nor does the mind exist as separated from the world. The dynamics of reversible and irreversible processes in a recursive-hierarchical structure is the only explanation that does justice to the intricate and unique relation between mind and the world. In other words, we constitute our separate and autonomous existence as long as we enact a given world based on our biological characteristics and the way they constitute the recursive-hierarchy.

In this context, semiosis plays a crucial role. Let me explain. Signs concern generalities communicated across domains. This point was emphasized in this book again and again. A sign is not the mental correspondence of a particular entity. When I say: "The cat is sitting on the chair" I may refer to a particular cat through the "the" but my use of the sign *cat* is the use of the set "cat". If signs are generalities then the inevitable question is how these generalities come into existence. More specifically, my question is whether we can speculate on the existence of a mechanism for producing generalities. My suggestion is to consider the emergence of signs and signification in terms of dimensionality reduction.

My first observation is that signification exists only in machines that actively enact a world. Again this observation should not be mistaken for a naïve representational theory assuming a simple and passive correspondence between the world and the mind. From the quantum level all the way up representation is active. It is enactment of the given world rather than passive mirroring.

This enactment necessarily involves dimensionality reduction. The idea of collapsing a higher-dimensional structure into a string of differences and repetitions is at the heart of the digital code that underlies semiosis. For example, the meaning of a word can be represented by a vector in a high-dimensional, semantic space. Practically, when communicated the word appears in a 1-D string of words and the addressee has to extract the meaning of the word from this low-dimensional representation.

As was realized by the psychoanalyst, Matte-Blanco (1988), dimensionality reduction entails *repetition*. For example, repetition exists when we map a 3-D structure into a 1-D string. Therefore, whenever repetition is evident, the existence of a higher-dimensional structure can be abduced. Dimensionality reduction is reduction in complexity but it has to be compensated for the inevitable loss of information. As the physics of computation teaches us, the loss of information is inevitable.

Repetition concerns non-substituted singularities (Deleuze, 1994, p. 1) but it is exactly these repetitions of singularities that are responsible for the emergence of generalities. They comprise the basic ingredient required for turning from pure differences to a difference that makes a difference—generality. Generality involves substitutes, values that underlie the analogue code: "generality expresses a point of view according to which one term may be exhausted or substituted for another" (p. 1). Generality results from repetition. Similarity and degree also result from repetition: "Gabriel Tarde suggested in this sense that resemblance itself was only *displaced repetition*" (p. 25; emphasis mine). The repetition results in points uniquely defined on the *n*-dimensional space being unique AND similar in the 1-D space. The repetition of A in Fig. 17.3 means that it is unique and not unique at the same time. This antinomy was realized by Matte-Blanco (1988) and I consider it to be the source of abstraction. Realizing the similarity (all A's are A) and the difference of elements (A's are different since they are located in different distances from the origin) is the source of sets = generalities. Repetition as a result of dimensionality reduction is at the heart of abstraction and semiosis, differences of similarities, and similarities of differences.

The concept of the sign demands more than difference and repetition. The concept of the sign concerns value that entails exchange and substitute. *Cat* is a sign as long as it can be exchanged for the set of yawning creatures. A $100 is a sign as long as it can be exchanged for a variety of commodities. The sign is not a literal substitute. The map is never the territory and the sign *cat* never yawns. Therefore, a sign has always a paradoxical nature referring and deflecting at the same time. Understanding the emergence of signs should begin with understanding the nature of a paradox. A paradox is actually one form of repetition in which the system oscillates between two diametrically opposed values without being able to settle on one of them. To quote Deleuze (1994):

> The real opposition [such as between 1 and 0 that define the bit] is not a maximum of difference [since a maximum exists along a continuum] but a minimum of repetition—a repetition reduced to two, echoing and returning on itself, a repetition which has the means to define itself. (p. 13)

In other words, a paradox is a re-entering form, a re-entering repetition. This re-entering repetition, like the paradoxical particle, which is in a state of superposition, is the origin of the bit. In other words, a paradox is at the heart of our digital code.

Repetition does not have to appear in a representational context. Repetition, as defined by Deleuze is a "difference without a concept". Music illustrates repetition that lacks any representational or propositional content. In fact the power of repetition in music is in its non-conceptual nature. In this sense, DNA is closer to musical chords than to the syntax of language since it is clearly a non-semantic structure. This idea resonates with Noble's musical metaphor of biological systems (Noble, 2006) and with the systems perspective in biology.

Poetry is also a good case for illustrating repetition but also a case that explains how meaning emerges from repetition. As suggested by Ezra Pound, "poetry is a language pared down to its essentials" and the essentials are differences and repetitions. Meaning emerges when these essentials are woven into a fabric.

In this context, we should remember that poetry has also an aspect of memorization and oblivion, two issues that repeatedly occupied me in this book. As suggested by Robert Frost, "poetry is what gets lost in translation". Using terminology introduced earlier, we can say that poetry is singularity that failed to be restored.

Memorization (reversibility) is deeply connected with repetition. We repeat, not only to strengthen neural connections in our brain, as suggested by the oversimplified cognitive theories but, in order to have *cues* for restoring the depth of high dimensionality. In this sense and as suggested by Mark Dotry: "Poetry is the physical enactment of a process of knowing by means of language". We know if we memorize and we memorize through repetition. DNA resonates with this logic. This is a new perspective on DNA. DNA is a 1-D string of differences and repetition that echoes the past. DNA is not a string of information; it is a string of *cues* being used by the organism for its own construction and maintenance. Our bodies know through differences and repetitions that echo our natural history as biological creatures and our memorable and delicate interactions with the environment. This is the Vedic perspective.

In Javanese culture, knowledge was not considered knowledge until it could be shaped into poetic form (Becker, 2000, p. 338). It may be the time to bring this wisdom to our understanding of living systems and to move on from mechanics to poiesis.

Cat-logue 5

Bamba:	Do you know the expression "curiosity killed the cat?"
Dr. N:	Sure. Why are you asking?
Bamba:	Lucky you are not a cat because you are shoving your nose into every possible subject from Indian poetry to molecular biology. Don't you want to be an expert?
Dr. N:	Someone said once that the expert is the enemy of democracy. The expert focuses only on his limited field of expertise and dissociates knowledge, unnaturally of course, from its systemic aspect. Although we would like our physician to be an *expert* in the sense of knowing the best he can about his field, we do not want him to dissociate knowledge.
Bamba:	You mean from context.
Dr. N:	Definitely.
Bamba:	The expert is the enemy of democracy exactly because he does not realize that "There is no *alibi* in existence". Philosophizing, or doing any scientific work without considering the consequences of our representations for being in the world might result in decadence.
Dr. N:	Nice to see that you are ending this book with an ethical lesson.
Bamba:	Both Bateson and Bakhtin pointed to the unavoidable link between epistemology and ethics. Between the way we conceive the world and the appropriate way of being in the world. It is comfortable to conceive organisms as marionettes of their selfish genes or as some kind of bio-physical toys produced by natural selection. This is the easiest way of ignoring our responsibility for the habitat in which we live. Bateson who was terrified by the destructive interference of man in nature and by the disastrous consequences of nuclear weapons could not accept the role of the scientist as a detached analytic philosopher, observing and commenting on a world to which he is indifferent.

Dr. N: In this sense he clearly realized that there is no *alibi* in existence and that wisdom, unless grounded in appropriate deeds, cannot be considered wisdom at all, even if it was published in *Science*.

Bamba: Correct, and let me illustrate this point with a Talmudic teaching from *Pirke Aboth*—The Saying of the Fathers—one of Judaism's most interesting ethical teachings.

Dr. N: A cat reading the Talmud? I am shocked!

Bamba: Please save me from your prejudices and listen. In Pirke Aboth (Herford, 1978, 3/22, p. 92) it is said:

> He [Rabbi Eleazr b. Azariah] used to say: "One whose wisdom is greater than his deeds what is he like? A tree whose branches are many and its roots few. And the wind comes and roots it up and overturns it on its face ..."

Dr. N: No *alibi* in existence.

Bamba: Not even for the wise guys.

References

Abbas, A.K., Lichtman, A.H. and Pober, J.S., 2000, *Cellular and Molecular Immunology*, W. B. Saunders, New York.

Akbar, A.N., Beverley, P.C. and Salmon, M., 2004, Will telomere erosion lead to a loss of T-cell memory? *Nat. Rev. Immunol.* 4, 737–743.

Alberts, B., Bray, D., Johnson, A., Lewis, J., Raff, M., Roberts, K. and Walter, P., 1998, *Essential Cell Biology*, Garland Publishing, New York.

Algom, D., 1986, *Perception and Psychophysics: A Historically Oriented Introduction to Some Representative Problems*, Ministry of Defense Publication, Tel Aviv.

Allegrucci, C., Thurston, A., Lucas, E. and Young, L., 2005, Epigenetics and the germline, *Reproduction* 129, 137–149. doi:10.1530/rep.1.00360.

Allende, I., 1992, *The Stories of Eva Luna*, Bantam Books, New York.

Anderson, C.C. and Matzinger, P., 2000, Danger: the view from the bottom of the cliff, *Semin. Immunol.* 12, 231–238.

Antonio Filippini, A., Riccioli, A., Padula, F., Lauretti, P., D'Alessio, A., De Cesaris, P., Gandini, L., Lenzi, A. and Ziparo, E., 2001, Immunology and immunopathology of the male genital tract: control and impairment of immune privilege in the testis and in semen, *Human Reprod. Update* 7, 444–449.

Aspinall, R., 2000, Longevity and immune response, *Biogerontology* 1, 272–278.

Baker, M.C., 2003, Syntax. In: *The Handbook of Linguistics* (M. Aronoff and J. Rees-Miller, Eds.), Blackwell Publishing, Oxford, 265–294.

Bakhtin, M.M., 1990, Author and hero in aesthetic activity. In: *Art and Answerability* (M. Holquist and V. Liapunov, Eds.), University of Texas Press, Austin, 4–257.

Bakhtin, M.M. 1999, In: *Toward a Philosophy of the Act* (V. Liapunov, Trans.), University of Texas Press, Austin.

Barbieri, M., 2002, *The Organic Codes: An Introduction to Semantic Biology*, Cambridge University Press, Cambridge.

Barbieri, M., 2004, The definition of information and meaning: two possible boundaries between physics and biology, *Riv. Biol.* 97, 91–110.

Barbieri, M., 2005, Life is "artifact-making", *J. Biosem.* 1, 113–142.

Barthes, R., 1977, *Music-Image-Text*, Fontana Collins, Glasgow.

Bateson, G., 1973, *Steps to an Ecology of Mind*, Granada Publishing, London.

Bateson, G., 1979, *Mind and Nature*, E. P. Dutton, New York.

Bateson, G., 2000, *Steps to an Ecology of Mind*, University of Chicago Press, Chicago.

Baulcombe, D., 2005, siRNA—the "dark matter" of genetics, *Pflanzenschutz-Nachrichten Bayer* 58, 21–33.

Bayne, C.J., 1990, Phagocytosis and non-self recognition in invertebrates, *BioScience* 40, 723–729.

Becker, A.L., 2000, *Beyond Translation: Essays Toward a Modern Philology*, University of Michigan Press, Ann Arbor.

Ben-Jacob, E., 1998, Bacterial wisdom, Gödel's theorem and creative genomic webs, *Physica A* 248, 57–76.

Bennett, C.H., 1982, The thermodynamics of computation—a review, *Int. J. Theor. Phys.* 21, 905–940.

Bergareche, A.M. and Ruiz-Mirazo, K., 1999, Metabolism and the problem of its universalization, *BioSystems* 49, 45–61.

Bergson, H.L., 1911, *Creative Evolution*, Macmillan, London.

Bergson, H.L., 1946, *The Creative Mind*, Wisdom Library, New York.

Bersini, H., 2003, Revising idiotypic immune networks, *Lect. Notes Comput. Sci.* 2801, 164–174.

Blundell, T.L. and Srinivasan, N., 1996, Symmetry, stability, and dynamics of multidomain and multicomponent protein systems, *PNAS* 93, 14243–14248.

Boden, M.A., 1999, Is metabolism necessary? *Brit. J. Phil. Sci.* 50, 231–248.

Bohm, D., 1998, In: *On creativity* (L. Nichol, Ed.), Routledge, New York.

Boniolo, G., 2003, Biology without information, *Hist. Philos. Life Sci.* 25, 257–275.

Borges, J.L., 1962, *Ficciones*, Grove Press, New York.

Borges, J.L., 2000a, In: *Collected Fictions* (A. Hurley, Trans.), Penguin Books, London.

Borges, J.L., 2000b, Ezra Pound as translator: between matter and form, *Hopscotch: A Cult. Rev.* 2 (1), 50–51.

Branden, C. and Tooze, J., 1999, *Introduction to Protein Structure*, Garland Publishing, New York.

Brown, J., 2000, *The Quest for the Quantum Computer*, Simon & Schuster, New York.

Buckland, J., 2002, Sex is good for you!, *Nat. Rev. Immunol.* 2, 148.

Burke, P., 2002, Context in context, *Common Knowl.* 8, 152–177.

Callard, R., George, A.J. and Stark, J., 1999, Cytokines, chaos, and complexity, *Immunity* 11, 507–513.

Carneiro, J. and Stewart, J., 1994, Rethinking "Shape Space": evidence from simulated docking suggests that steric shape complementarity is not limiting for antibody–antigen recognition and idiotypic interactions, *J. Theor. Biol.* 169, 391–402.

Carnie, A., 2002, *Syntax: A Generative Introduction*, Blackwell Publishing, Oxford.

Chan, S.W.L., Zilberman, D., Xie, Z., Johansen, L.K., Carrington, J.C. and Jacobsen, S.E., 2004, RNA silencing genes control de novo DNA methylation, *Science* 303 (5662), 1336.

Chomsky, N., 1957, *Syntactic Structures*, Mouton, The Hague.

Chomsky, N., 1965, *Aspects of the Theory of Syntax*, MIT Press, Cambridge, MA.

Chomsky, N., 1975, *The Logical Structure of Linguistic Theory*, Plenum Press, New York.

Choudhury, S.R.C. and Knapp, L., 2000, Human reproductive failure II: immunogenetic and interacting factors, *Hum. Reprod. Update* 7, 113–134.

Clardy, J., 1999, Borrowing to make ends meet, *PNAS* 96, 1826–1827.

Clegg, J.S., 2001, Cryptobiosis—a peculiar state of biological organization, *Comp. Biochem. Physiol. B* 128, 613–624.

Cleland, C.E., 2004, The concept of computability, *Theor. Comput. Sci.* 317, 209–225.

Cleland, C.E. and Chyba, C.F., 2002, Defining 'life', *Orig. Life Evol. Biosph.* 32 (4), 387–393.

Cogoni, C. and Macino, G., 2000, Post-transcriptional gene silencing across kingdoms, *Curr. Opin. Genet. Dev.* 10, 638–643.

Cohen, I., 1992, The cognitive principal challenges clonal selection, *Immunol. Today* 13, 490.

Cohen, I.R., 1994, Kadishman's tree, Escher's angels, and the immunological homunculus. In: *Autoimmunity: Physiology and Disease* (A. Coutinho and M.D. Kazatchkine, Eds.), Wiley-Liss Inc., New York, 7–18.

Cohen, I.R., 2000a, *Tending Adam's Garden*, Academic Press, New York.

Cohen, I.R., 2000b, Discrimination and dialogue in the immune system, *Semin. Immunol.* 12, 21–219.

Cohen, I., 2001, Immunity, set points, reactive systems, and allograft rejection, *Graft* 4, 365–382.

Cohen, I.R., 2006, Information landscapes in art, science, and evolution, *Bull. Math. Biol.* 68, 1213–1229.

Cohen, I.R., Hershberg, U. and Solomon, S., 2004, Antigen-receptor degeneracy and immunological paradigm, *Mol. Immunol.* 40, 993–996.

Conrad, M., 1996, Cross-scale information processing in evolution, development and intelligence, *BioSystems* 38, 97–109.

Conrad, P.G. and Nealson, K.H., 2001, A non-earthcentric approach to life detection, *Astrobiology* 1, 15–24.

Couzin, J., 2002, Breakthrough of the year: small RNAs make big slash, *Science* 298, 2296–2297.

CP, Collected papers. *Collected Papers of Charles Sanders Peirce* (Vols. 1–8), Harvard University Press, Cambridge, MA. (Vols. 1–6, C. Hartshorne & P. Weiss, Eds.; Vols. 7–8, A. W. Burks, Ed.).

Crist, E. and Tauber, A.I., 1999, Selfhood, immunity and the biological imagination: the thought of Frank Macfarlane Burnet, *Biol. Philos.* 15, 509–533.

Cropley, D.H., 1998, Toward formulating a semiotic theory of measurement information—Part 1, *Measurement* 24, 237–248.

Cvrckova, F. and Markos, A., 2005, Beyond bioinformatics: Can similarity be measured in the digital world? *J. Biosem.* 1, 93–112.

Darwin, C., 1958, *The Origin of Species*, Mentor Books, New York.

Dawkins, R., 1976, *The Selfish Gene*, Oxford University Press, New York.

De Cesaris, P., Filippini, A., Cervelli, C., Riccioli, A., Muci, S., Starace, G., Stefanini, M. and Ziparo, E., 1992, Immunosuppressive molecules produced by

Sertoli cells cultured in vitro: biological effects on lymphocytes, *Biochem. Biophys. Res. Commun.* 186, 1639–1646.

Deleuze, G., 1990, In: *The Logic of Sense* (M. Lester, Trans.), Columbia University Press, New York.

Deleuze, G., 1994, In: *Difference and Repetition* (P. Patton, Trans.), Columbia University Press, New York.

Denbigh, K.G., 1989, Note on entropy, disorder, and dis-organization, *Brit. J. Phil. Sci.* 40, 323–333.

Deutsch, D., Ekert, A. and Lupacchini, R., 2000, Machines, logic and quantum physics, *Bull. Symbolic Logic* 6, 265–283.

DeVeale, B., Brummel, T. and Seroude, L., 2004, Immunity and aging: the enemy within? *Aging Cell* 3, 195–208.

Dill, K.A. and Bromberg, S., 2003, *Molecular Driving Forces: Statistical Thermodynamics in Chemistry and Biology*, Garland Science, New York.

Dong, S. and Searls, D.B., 1994, Gene structure prediction by linguistic methods, *Genomics* 23, 540–551.

Downward, J., 2004, RNA interference, *BMJ* 328, 1245–1248.

Dufour, J.M., Gores, P., Hemendinger, R., Emerich, D.F. and Halberstadt, C.R., 2004, Transgenic Sertoli cells as a vehicle for gene therapy, *Cell Transplant.* 13, 1–6.

Dupre, J., 1993, *The Disorder of Things*, Harvard University Press, Cambridge, MA.

Eddy, S.R., 1999, Noncoding RNA genes, *Curr. Opin. Genet. Dev.* 9, 695–699.

Eddy, S.R., 2001, Non-coding RNA genes and the modern RNA world, *Nat. Rev. Genet.* 2, 919–929.

Efroni, S. and Cohen, I., 2002, Simplicity belies a complex system: a response to the minimal model of immunity of Langman and Cohn, *Cell Immunol.* 216, 23–30.

Efroni, S. and Cohen, I.R., 2003, The heuristics of biologic theory: the case of self-noself discrimination, *Cell Immunol.* 223, 87–89.

Efroni, S., Harel, D. and Cohen, I., 2003, Toward rigorous comprehension of biological complexity: Modeling execution, and visualization of Thymic T-Cell maturation, *Genome Research* 13, 2485–2497.

Emmeche, C. and Hoffmeyer, J., 1991, From language to nature, *Semiotica* 84, 1–42.

Emmeche, C., Køppe, S. and Stjernfelt, F., 2000, Levels of emergence, and three versions of downward causation. In: *Downward Causation: Minds, Bodies and Matter* (P.B. Andersen, C. Emmeche, N.O. Finnemann and P.V. Christiansen, Eds.), Aarhus University Press, Arhus, 13–34.

EP, The essential Peirce. *The Essential Peirce. Selected Philosophical Writings* (Vols. 1–2), Indiana University Press, Bloomington, IN. (Vol. 1 [1867–1893], N. Houser and C. Kloesel, Eds., 1992; Vol. 2 [1893–1913], Peirce Edition Project, Ed., 1998).

Feofilova, E.P., 2003, Deceleration of vital activity as a universal biochemical mechanism ensuring adaptation of microorganisms to stress factors: a review, *Appl. Biochem. Microbiol.* 39 (1), 1–18.

Ferguson, J., 1976, *An Illustrated Encyclopedia of Mysticism and Mystery Religions*, Thames and Hudson, London.

Finkelstein, A.V. and Galzitskaya, O.V., 2004, Physics of protein folding, *Phys. Life Rev.* 1, 23–56.

Fleck, L., 1979, *Genesis and Development of a Scientific Fact*, University of Chicago Press, Chicago.

Franceschi, C., Bonafè, M. and Valensin, S., 2000, Human immunosenescence: the prevailing of innate immunity, the failing of clonotypic immunity, and the filling of immunological space, *Vaccine* 18, 1717–1720.

Frank, M.P., 2002, The physical limits of computing, *Comput. Sci. Eng.* 4, 16–26.

Gardner, M., 1996, *The Night is Large*, St. Martin's Press, New York.

Genesis Rabbah: The Judaic Commentary to the Book of Genesis, 1985, (J. Neusner, Trans.), Scholars Press (Brown Judaic Studies), Atlanta, GA.

Gentner, D., 1983, Structure-mapping: a theoretical framework for analogy, *Cognit. Sci.* 7, 155–170.

Gilbert, S.F., 2000, *Developmental Biology*, 6th ed., Sinauer Associates, Sunderland, MA.

Gilbert, S.F., 2001, Ecological developmental biology: developmental biology meets the real world, *Dev. Biol.* 233, 1–12.

Gilbert, S.F., 2005, Mechanism for the environmental regulation of gene expression: ecological aspects of animal development, *J. Biosci.* 30, 65–74.

Godfrey-Smith, P., 2002, On genetic information and genetic coding. In: *In the Scope of Logic, Methodology, and the Philosophy of Science* (P. Gardenfors, J. Wolenski and K. Kajania-Placek, Eds.), Kluwer, Dordrecht, Vol. 2, 387–400.

Goodwin, B., 1994, *How the Leopard Changed Its Spots*, Weidenfeld and Nicolson, London.

Griswold, M.D., 1995, Interactions between germ cells and Sertoli cells in the testis, *Biol. Reprod.* 52, 211–216.

Guo, S. and Kemphues, K.J., 1995, Par-1, a gene required for establishing polarity in C. elegans embryos, encodes a putative Ser/Thr kinase that is asymmetrically distributed, *Cell* 81, 611–620.

Guppy, M., 2004, The biochemistry of metabolic depression: a history of perceptions, *Comp. Biochem. Physiol. Part B* 139, 435–442.

Gupta, S. Su, H., Bi, R., Agrawal, S. Gollapudi, S., 2005, Life and death of lymphocytes: a role in the immunesenescene, *Immun. Aging* 2, 212. doi: 10.1186/1742-4933-2-12.

Guyton, A.C. and Hall, J.E., 1996, *Textbook of Medical Physiology*, 9th ed., W. B. Saunders Company, Philadelphia.

Hammond, S.M., Caudy, A.A. and Hannon, G.J., 2001, Post-transcriptional gene silencing by double-stranded RNA, *Nat. Rev. Gen.* 2, 110–119.

Harries-Jones, P., 1995, *A Recursive Vision: Ecological Understanding and Gregory Bateson*, Toronto University Press, Toronto.

Harries-Jones, P., 2006, *Re-Generating Environment: The Logic of Moving Boundaries*. Unpublished manuscript.

Harris, Z.S., 1968, *Mathematical Structures of Language*, Interscience, New York.

Harris, Z.S., 1991, *A Theory of Language and Information: A Mathematical Approach*, Clarendon Press, Oxford.

Herford, R.T. (Ed. and Trans.), 1978, *Prike Aboth*, Shoken Books, New York.

Hewitt, P.G., 1993, *Conceptual Physics*, HarperCollins College Publishers, New York.

Hoey, M., 2005, *Lexical Priming: A New Theory of Words and Language*, Routledge, New York.

Hoffmeyer, J., 1996, *Signs of Meaning in the Universe*, Indiana University Press, Bloomington, IN.

Hoffmeyer, J. and Emmeche, C., 1991, Code-duality and the semiotics of nature. In: *Semiotic Modeling* (M. Anderson and F. Merrell, Eds.), Mouton de Gruyter, New York, 117–166.

Holquist, M., 1990a, *Dialogism*, Routledge, London.

Holquist, M., 1990b, Introduction, In: *Art and Answerability* (M.M. Bakhtin, Ed.), University of Texas Press, Austin.

Holquist, M., 1999, Foreword, In: *Toward a Philosophy of the Act* (M.M. Bakhtin, Ed.), University of Texas Press, Austin, vii–xv.

Howes, M., 1998, The self of philosophy and the self of immunology, *Perspect. Biol. Med.* 42, 118.

Huxley, A., 1931, *Music at Night*, Chatto & Windus, London.

Jablonka, E., 2002, Information: its interpretation, its inheritance, and its sharing, *Philos. Sci.* 69, 578–605.

Jackendoff, R., 2003, *Foundations of Language*, Oxford University Press, New York.

Jakobson, R., 1971, *Selected Writings*, Mouton, The Hague.

Janeway, C.A., Jr., 1992, The immune system evolved to discriminate infectious nonself from noninfectious self, *Immunol. Today* 13, 6–11.

Jaworski, A., 1997, Introduction: an overview. In: *Silence: Interdisciplinary Perspective* (A. Jaworski, Ed.), Mouton de Gruyter, Berlin, 3–15.

Jerne, N.K., 1974, Toward a network theory of the immune system, *Ann. Immunol. (Inst. Pasteur.)* 125c, 373–389.

Jerne, N.K., 1984, Idiotypic networks and other preconceived ideas, *Immunol. Rev.* 79, 5–24.

Ji, S., 1997, Isomorphism between cell and human languages: molecular biological, bioinformatics and linguistic implications, *BioSystems* 44, 17–39.

Jones, 1995, *Language of the Genes*, Anchor, New York.

Jones, P.A. and Takai, D., 2001, The role of DNA methylation in mammalian epigenetics, *Science* 293 (5532), 1068–1070.

Kanehisa, M., 2000, *Post-Genome Informatics*, Cambridge University Press, Cambridge.

Kass, R.E., 1994, Editor's remarks in WRR94, *Stat. Sci.* 9, 306.

Kass, R.E., 1999, Introduction to "Solving the Bible Code Puzzle" by Brendan McKay, Dror Bar-Natan, Maya Bar-Hiller and Gil Kalai, *Stat. Sci.* 14, 149.

Keller, E.-F., 2000, *The Century of the Gene*, Harvard University Press, Cambridge, MA.

Kempe, B., 1886, A memoir on the theory of mathematical form, *Philos. Trans. R. Soc. London* 177, 1–70.

Kidwell, M.G. and Lisch, D.R., 2001, Perspective: transposable elements, parasitic DNA, and genome evolution, *Evolution* 55, 1–24.

Koessi, J., Salminen, P., Rantala, A. and Laato, M., 2003, Population-based study of the surgical workload and economic impact of bowel obstruction caused by postoperative adhesions, *Brit. J. Surg.* 90, 1441–1444.

Koshland, D.E., 2002, The seven pillars of life, *Science* 295, 2215–2216.

Lakoff, G. and Johnson, M., 1999, *Philosophy in the Flesh*, Basic Books, New York.

Landauer, R., 1961, Irreversibility and heat generation in the computing process, *IBM J. Res. Dev.* 5, 183–191.

Landauer, R. and Bennett, C., 1985, The fundamental physical limits of computation, *Sci. Am.* 253, 48–56.

Lande-Diner, L. and Cedar, H., 2005, Silence of the genes—mechanism of long-term repression, *Nat. Rev. Genet.* 6, 648–654.

Lange, M., 1996, Life, "artificial life" and scientific explanation, *Philos. Sci.* 63 (2), 225–244.

Langman, R.E. and Cohn, M., 2000, A minimal model for the self-nonself discrimination: a return to the basics, *Semin. Immunol.* 12, 180–195.

Laughlin, R.B., Pines, D., Schmalian, J., Stojkovic, B.P. and Wolynes, P., 2000, The middle way, *Proc. Natl. Acad. Sci. U.S.A.* (97), 32–37.

Leff, H.S. and Rex, A.F. (Eds.), 1990, *Maxwell's Demon: Entropy, Information, Computing*. IOP Publishing, Bristol.

Lewontin, R.C., 1991, *Biology as Ideology*, Harper Perennial, New York.

Luigi-Luisi, P., 1998, About various definitions of life, *Orig. Life Evol. Biosph.* (28), 613–622.

Luria, A.R. and Vygotsky, L.S., 1992, *Ape, Primitive Man and Child: Essays in the History of Behavior*, Paul M. Deutsch Press, Orlando, FLA.

Mackay, C.R. and von Andrian, U.H., 2001, Memory T cells—local heroes in the struggle for immunity, *Science* 291, 2323–2324.

MacLeod, A.D. and Maycock, E., 1992, Awareness during anesthesia and post traumatic stress disorder, *Anesth. Intern. Care* 20, 378–382.

Makalowski, W., 2003, Not junk after all, *Science* 300, 1246–1247.

Manning, C.D. and Schutze, H., 2003, *Foundations of Statistical Natural Language Processing*, MIT Press, Cambridge, MA.

Mapara, M.Y. and Sykes, M., 2004, Tolerance and cancer: mechanisms of tumor evasion and strategies for breaking tolerance, *J. Clin. Oncol.* 6, 1136–1151.

Marchalonis, J.J. and Schluter, S.F., 1990, Origins of immunoglobulins and immune recognition molecules, *BioScience* 40, 758.

Markos, A., 2002, *Readers of the Book of Life*, Oxford University Press, Oxford.

Matte-Blanco, I., 1988, *Thinking, Feeling and Being*, Routledge, London.

Mattick, J.S., 2001, Non-coding RNAs: the architects of eukaryotic complexity, *EMBO Rep.* 2 (11), 986–991.

Mattick, J.S., 2003, Challenging the dogma: the hidden layer of non-protein-coding RNAS in complex organism, *BioEssays* 25, 930–939.

Mattick, J.S. and Gagen, M.J., 2001, The evolution of controlled multitasked gene networks: the role of introns and other noncoding RNAs in the development of complex organism, *Mol. Biol. Evol.* 18 (9), 1611–1630.

Maturana, H.R. and Varela, F.J., 1992, *The Tree of Knowledge: The Biological Roots of Human Understanding*, Shambhala, Boston.

Matzinger, P., 2002, The danger model: a renewed sense of self, *Science* 296, 301–305.

Maynard Smith, J., 2000, The concept of information in biology, *Philos. Sci.* 67, 177–194.

McKay, B., Bar-Natan, D., Bar-Hillel, M. and Kalai, G., 1999, Solving the bible code puzzle, *Stat. Sci.* 14, 150–173.

McLachlan, R., 2002, Basis, diagnosis and treatment of immunological infertility in men, *J. Reprod. Immunol.* 57, 35.

Meehan, R.R., 2003, DNA methylation in animal development, *Semin. Cell Dev. Biol.* 14, 53–65.

Mellor, A.L. and Munn, D.H., 2000, Immunology at the maternal-fetal interface: lessons for the T cell tolerance and suppression, *Annu. Rev. Immunol.* 18, 367–391.

Merrell, F., 1997, *Peirce, Signs and Meaning*, Toronto University Press, Toronto.

Mey, J.L., 1998, When voices clash: A study in literary pragmatics. *Trends in Linguistics. Studies and Monographs 115*. Mouton de Gruyter, Berlin.

Mey, J.L., 2001, *Pragmatics: An Introduction*, Blackwell, Oxford.

Mill, J.S., 1911, *A System of Logic Ratiocinative and Inductive: Being a Connected View of the Principles of Evidence and the Methods of Scientific Investigation*, Longman, London.

Morson, G.S. and Emerson, C., 1990, *Mikhail Bakhtin: Creation and Prosaics*, Stanford University Press, Stanford.

Mruk, D.D. and Yan Cheng, C., 2004, Sertoli-Sertoli and Sertoli germ cell interactions and their significance in germ cell movement in the seminiferous epithelium during spermatogenesis, *Endocr. Rev.* doi: 10.1210/2003-2003.

Nagata, S., 1999, Fs ligand-induced apoptosis, *Annu. Rev. Genet.* 33, 29–55.

Nelson, D.R., 2002, Current status of the tardigrada: evolution and ecology, *Integr. Comp. Biol.* 42, 652–659.

NEM, *The New Elements of Mathematics*, by Charles S. Peirce. Four volumes in five books. Edited by Carolyn Eisele (1976). Mouton Publishers, The Hague.

Neuman, Y., 2003a, *Processes and Boundaries of the Mind: Extending the Limit Line*, Springer, New York.

Neuman, Y., 2003b, Mobius and paradox: on the abstract structures of boundary events in semiotic systems, *Semiotica* 147, 135–148.

Neuman, Y., 2004a, What does pain signify? A hypothesis concerning pain, the immune system and unconscious pain experience under general anesthesia, *Med. Hyp.* 63, 1051–1053.

Neuman, Y., 2004b, Meaning-making in the immune system, *Perspect. Biol. Med.* 47, 317–328.

Neuman, Y., 2005, The immune self, the sign and the testes, *SEED J.* 5, 85–109.

Neuman, Y., 2006, Cryptobiosis: a new theoretical perspective, *Prog. Biophys. Mol. Biol.* 92, 258–267.

Nielsen, M.A. and Chuang, I.L., 2000, *Quantum Computation and Quantum Information*, Cambridge University Press, Cambridge.

Nijhout, H.F., 1990, Metaphors and the role of genes in development, *BioEssays* 12 (9), 441–446.

Noble, D., 2006, *The Music of Life*, Oxford University Press, Oxford.

Nodder, S. and Martin, P., 1997, Wound healing in embryos: a review, *Anat. Emryol.* 195, 215–228.

Orosz, C.G., 2001, How complexity helps to shape alloimmunity, *Graft* 4, 1–3.

O'Shea, J.J., Ma, A. and Lipsky, P., 2001, Cytokines and autoimmunity, *Nat. Rev. Immunol.* 2, 37–45.

Pankanti, S., Prabhakar, S. and Jain, A.K., 2002, On the individuality of fingerprints, *Pattern Anal. Mach. Intell.* 24, 1010–1025.

Paton, R., 2002, Process, structure and context in relation to integrative biology, *BioSystems* 64, 63–72.

Pattee, H.H., 2001, The physics of symbols: bridging the epistemic cut, *BioSystems* 60, 5–21.

Pearson, W.R. and Wood, T.C., 2001, Statistical significance in biological sequence comparison. In: *Handbook of Statistical Genetics* (D.J. Balding, M. Bishop and C. Cannings, Eds.), John Wiley and Sons, New York, 39–62.

Peirce, C.S., 1955, The principles of phenomenology. In: *Philosophical Writings of Peirce* (J. Buchler, Ed.), Dover Publications, New York, 74–98.

Perelson, A., 1989, Immune network theory, *Immunol. Rev.* 110, 5–36.

Perkins, D.O., Jeffries, C. and Sullivan, P., 2005, Expanding the 'central dogma': the regulatory role of non-protein coding genes and implications for the genetic liability to schizophrenia, *Mol. Psychiatr.* 10, 69–78.

Piaget, J., 1970, *Structuralism*, Basic Books, New York.

Plenio, M.B. and Vitelli, V., 2001, The physics of forgetting: Landauer's erasure principle and information theory, *Contemp. Phys.* 42 (1), 25–60.

Polanyi, M., 1968, Life's irreducible structure, *Science* 160, 1308–1312.

Redd, M.J., Cooper, L., Wood, W., Stramer, B. and Martin, P., 2004, Wound healing and inflammation: embryos reveal the way to perfect repair, *Phil. Trans. R. Soc. London B* 359, 777–784.

Reik, W., Dean, W. and Walter, J., 2001, Epigenetic reprogramming in mammalian development, *Science* 293 (5532), 1089–1093.

Rosen, S., 1994, *Science, Paradox, and the Moebius Principle*, SUNY Press, New York.

Rosen, S.M., 2004, What is a radical recursion? *SEED J.* 4 (1), 38–57.

Roth, P., 1999, *I Married a Communist*, Vintage, London.

Rucker, R., 1977, *Geometry, Relativity and the Fourth Dimension*, Dover, New York.

Russell, B., 1903, *The Principles of Mathematics*, George Allen & Unwin, London.

Russell, B., 1908, Mathematical logic as based on the theory of types, *Am. J. Math.* 30 (3), 222–262.

Sampson, G., 1980, *Making Sense*, Oxford University Press, Oxford.

Sarkar, S., 1996, Biological information: a skeptical look at some central dogmas of molecular biology. In: *The Philosophy and History of Molecular Biology: New Perspectives* (S. Sarkar, Ed.), Kluwer Academic Publishers, Dordrecht, 187–231.

Saussure, F.de., 1959, *Course in General Linguistics*, McGraw-Hill, New York.

Saussure, F.de., 1972, In: *Course in General Linguistics* (R. Harris, Trans.), Duckworth, London.

Schill, R.O., Steinbruck, G.H.B. and Kohler, H.R., 2004, Stress gene (hsp70) sequences and quantitative expression in *Milensium tardigradum* (Tardigrade) during active and cryptobiotic stages, *J. Exp. Biol.* 207, 1607–1613.

Schwartz, M., Moalem, G., Leibowitz-Amit, R. and Cohen, I.R., 1999, Innate and adaptive immune responses can be beneficial for CNS repair, *Trends Neurosc.* 22, 295–299.

Searls, D.B., 2002, The language of the genes, *Nature* 420, 211–217.

Sebeok, T.A., 2001, Biosemiotics: its roots, proliferation, and prospects, *Semiotica* 134, 61–78.

Sercarz, E., Celada, F., Mitchison, N.A. and Tada, T. (Eds.), 1998, *The Semiotics of Cellular Communication in the Immune System*. Springer, New York.

Shahn, B., 1954, *The Alphabet of Creation: An Ancient Legend from the Zohar*, Shocken Books, New York.

Shanon, B., 1992, Metaphor: from fixedness and selection to differentiation and creation, *Poetics Today* 13, 660–685.

Shanon, B., 1993, *The Representational and the Presentational: An Essay on Cognition and the Study of Mind*, Harvester-Wheatsheaf, London.

Shea, J.E., Onuchic, J.N. and Brooks, C.L., 2000, Energetic frustration and the nature of the transition state in protein folding, *J. Chem. Phys.* 113, 7663–7671.

Shukman, A. (Ed.), 1983, *Bakhtin School Papers*. RPT Publications, Oxford.

Shulman, D., 1986, *Prakim BaShira HaHodit* [*Lectures on Indian Poetry*], Ministry of Defense Publication, Tel-Aviv.

Shulman, S., 1995, Selection of 'antibody-free' spermatozoa from semen of patients with anti-sperm antibodies, *Hum. Reproduction* 10, 975–976.

Sible, 2003, Cell biology: thanks for the memory, *Nature* 426, 392.

Silverstein, A.M., 2002, The clonal selection theory: what is really is and why modern challenges are misplaced, *Nature* 3, 793–796.

Smyth, M.J., Cretney, E., Kershaw, M.H. and Hayakawa, Y., 2004, Cytokines in cancer immunity and immunotherapy, *Immunol. Rev.* 202, 275–293.

Snyder, M. and Gerstein, M., 2003, Genomics: defining genes in the genomic era, *Science* 300, 258–260.

Spencer-Brown, G., 1994, *Laws of Form*, Cognizer, Portland, OR.

Sprent, J. and Tough, D.F., 2001, T cell death and memory, *Science* 293, 245–247.

Storz, G., 2002, An expanding universe of noncoding RNAs, *Science* 296, 1260–1263.

Strohman, R.C., 2000, Organization becomes cause in the matter, *Nat. Biotechnol.* 18, 575–576.

Swift, J., 1994, *Gulliver's Travels*, Penguin, UK.

Tauber, A.I., 1996, *The Immune Self: Theory or Metaphor?* Cambridge University Press, Cambridge.

Tauber, A.I., 1997, Historical and philosophical perspectives concerning immune cognition, *J. Hist. Biol.* 30, 419–440.

Tauber, A.I., 1998, Conceptual shifts in immunology: comments on the "two-way" paradigm, *Theor. Med. Bioethics* 19, 457–473.

Tauber, A. I., 2002, The biological notion of self and non-self. *Stanford Encyclopedia of Philosophy*. Available at http://plato.stanford.edu/entries/biology-self.

The Economist, 1997, How (and why) to find a needle in a haystack, April 5–11, pp. 105–107 (British Edition).

The MIT Guide to Lock Picking, 1991, *Theodore. T. Tool.* Available at http://www.toool.nl/mit.pdf

Thibault, P.J., 2005, *Saussure and Beyond. Lecture One: Speech and Writing: Two Distinct Systems of Sign.* Available at www.chass.utoronto.ca/epc/srb/cyber/thi1.html

Thomas, D.E., 1997, Hidden messages and the Bible code, *Skeptical Inquirer* 21 (6), 30–36.

Todaro, M., Zeuner, A. and Stassi, G., 2004, Role of apoptosis in autoimmunity, *J. Clin. Immunol.* 24, 1–11.

Tung, K.S.K., 1980, Autoimmunity in the testis. In: *Immunological Aspects of Infertility and Fertility Regulation* (D. Dhindsa and G.F.B. Shumacher, Eds.), Elsevier, New York, 33–91.

Vakkila, J. and Lotze, M.T., 2004, Inflammation and necrosis promote tumor growth, *Nat. Rev. Immunol.* 4, 641–648.

Vance, R.E., 2000, A Copernican revolution? Doubts about the danger theory, *J. Immunol.* 165, 1725–1728.

Van Speybroeck, L., van de Vijver, G. and de Waele, D., 2002, From epigenesis to epigenetics: the genome in context, *Ann. NY Acad. Sci.* 981.

Vasto, S., Malavolta, M. and Pawelec, G., 2006, Age and immunity, *Immun. Aging* 3, 2. doi: 10.1186/1742-4933-3-2.

Vaz, N.M. and Varela, F.J., 1978, Self and non-sense: an organism-centered approach to immunology, *Med. Hyp.* 4, 231–267.

Voinnet, O., 2002, RNA silencing: small RNAs as ubiquitous regulators of gene expression, *Curr. Opin. Plant Biol.* 5, 444–451.

Volosinov, V.N., 1986, *Marxism and the Philosophy of Language*, Harvard University Press, Cambridge, MA.

von Foerster, H. and Poerksen, B., 2002, *Understanding Systems: Conversations on Epistemology and Ethics*, Kluwer Academic/Plenum Publishers, New York.

von Uexküll, J., 1982, The theory of meaning, *Semiotica* 42, 25–82.

Waddington, C.H., 1957, *The Strategy of the Genes*, George Allen and Unwin, London.

Wade, M.J., Winther, R.G., Agrawal, A.F. and Goodnight, C.J., 2001, Alternative definitions of epistasis: dependence and interaction, *Trends Ecol. Evol.* 16, 498–504.

Wagner, R.P., Maguire, M.P. and Stallings, R.L., 1993, *Chromosomes—A Synthesis*, Wiley-Liss, New York.

Wagner, S.D. and Neuberger, M.S., 1996, Somatic hypermutation of immunoglobulin genes, *Annu. Rev. Immunol.* 14, 441–457.

Waldman, H. and Cobbold, S., 2004, Exploiting tolerance processes in transplantation, *Science* 305, 209–212.

Wassenegger, M., 2000, RNA-directed DNA methylation, *Plant Mol. Biol.* 43, 203–220.

Watanabe, M., Kikawada, T., Minagawa, N., Yukuhiro, F. and Okuda, T., 2002, Mechanism allowing an insect to survive complete dehydration and extreme temperatures, *J. Exp. Biol.* 205, 2799–2802.

Wegner, P., 1998, Interactive foundations of computing, *Theor. Comput. Sci.* 192, 315–351.

Weyl, H., 1957, *Symmetry*, Princeton University Press, Princeton, NJ.

Witztum, D., Rips, E. and Rosenberg, Y., 1994, Equidistant letter sequences in the book of Genesis, *Stat. Sci.* 9, 429–438.

Woolley, K. and Martin, P., 2000, Conserved mechanisms of repair: from damaged single cells to wounds in multicellular tissues, *BioEssays* 22, 911–919.

Wright, J.C., 2001, Cryptobiosis 300 years on from van Leuwenhoek: What have we learned about tardigrades? *Zool. Anz.* 240, 563–582.

Xiong, and Ferrell, 2003, A positive-feedback-based bistable 'memory module' that governs a cell fate decision, *Nature* 426, 460–465.

Yates, F.E., 1993, Self-organizing systems. In: *The Logic of Life* (C.A.R. Boyd and D. Noble, Eds.), Oxford University Press, Oxford, 189–219.

Zimmerman, C.M. and Padgett, R.W., 2000, Transforming growth factor beta signaling mediators and modulators, *Gene* 249, 17–30.

Žižek, S. and Daly, G., 2004, *Conversations with Zizek*, Cambridge University Press, Cambridge.

Name Index

Bakhtin, M., 39, 91, 140, 162–163, 181, 200, 239–241, 243, 245–255, 257–258, 272, 279
Barbieri, M., 48–49, 173, 176, 209, 229, 272
Bateson, G., 48–49, 51–53, 58, 62–63, 65–67, 86, 117–120, 123, 136, 143, 159–160, 168–169, 174–175, 186, 198, 200, 208, 210, 213, 215, 218–221, 245–250, 252, 256, 265, 279
Becker, A.L., 53, 166, 191–192, 230–231, 237, 240, 256, 269
Bergson, H., 7, 115–116, 119–120, 123, 126, 229, 278
Borges, J.L., 15, 49–50, 61–62, 66, 212–213, 231, 235, 264

Chomsky, N., 134, 148–149, 151, 153–154, 271, 279
Cohen, I., 3, 42, 56, 75, 79, 81, 84–85, 87, 89–93, 102, 112–116, 124, 136, 138, 140, 143–144, 176–177, 201, 222, 231, 234, 236, 238, 243, 265

Dawkins, R., 7, 13, 15, 53, 74, 189
Deleuze, G., 26, 46, 68, 228, 251, 267–268, 273

Fleck, L., 11–13

Gasset, J.O., 53, 191, 230, 256

Harries-Jones, P., 86, 143, 218–219
Harris, Z., 21, 23, 151–152
Hoffmeyer, J., 42, 51, 136, 175–176, 208, 272
Holquist, M., 41, 163, 239–241, 245, 250–251, 254–255, 258

Jerne, N.K., 3, 108–112, 120, 233

Keller, E.-F., 15–17, 39

Landauer, R., 45, 50, 58–59, 64, 223–224, 232
Laughlin, R.B., 141, 169, 221
Lewontin, R.C., 8–9, 15, 17

Markos, A., 53, 159, 229, 272
Mattick, J.S., 56–57, 60–61, 69
Matte-Blanco, I., 68, 210, 245, 267
Maturana, H., 21, 27, 110, 152, 192, 230
Mey, J., 135, 162–163, 166, 185–186

Neuman, Y., 47, 63, 201, 212, 218–219, 229, 231–233, 238, 263, 274, 277
Noble, D., 268, 271

Peirce, C.S., 3, 25, 47, 90–91, 108, 111–112, 120, 139–142, 279
Piaget, J., 89, 196–198, 279
Polanyi, M., 41, 44–45, 49, 137, 152, 221, 279

Rosen, S., 88, 111, 157, 227, 253
Russell, B., 26–28, 137

Saussure, F.de., 3, 21, 105–108, 111, 139, 144, 152, 209–210, 241, 252, 279
Spencer-Brown, G., 68, 252
Strohman, R.C., 169, 218, 221–222, 231, 246
Szymborska, W., 211, 264
Tauber, A.I., 55–56, 96–97, 101, 105, 109–110, 133, 229, 233
Volosinov, V., 3, 112, 116–117, 123–125, 127, 129, 144, 160, 162–166, 189, 210, 239–240, 245, 255, 257, 279

Subject Index

abductive inference/reasoning, 90
alphabet, 19
analogue, 51
antibody, 80
autoimmunity, 102

basins of attraction, 90
biosemiotics, 42
book of life, 9
bootstrapping, 220
boundary conditions, 41

Clonal Selection Theory, 106
code duality, 51
codon, 67
coin-flip gate, 205
computation, 45
constraints, 23
Context, xii
contextualist approach, 112
coordination-under-constraints, 177
co-respondence, 91
cryptobiosis, 215, 216
cytokines, 79

danger model, 98
death, 217
difference that makes a difference, 24
differences, 28
digital, 51
digital code, 23
dimension, 242
dimensionality reduction, 239
dogma, 12
downward causation, 220, 221

entropy, 159
enzymes, 35
epigenesis, 70

Firstness, 25
form, 119

gene silencing, 60
Genetics, 15
Genome, xii
Genome Project, 15
gestalt, xii
grammar, 21
Group theory, 197

habit, 90
hermeneutic circularity, 137
Hypothetical inference, 90

ideology, 8
imaginary number, 68
Immune memory, 232
Immune recognition, 79
Immune Self, 95, 97, 99, 101, 103, 105, 107, 109, 111, 113, 115, 117, 119, 121, 123, 125, 127
Immune Specificity, 83
in between, xx, 118
inflammation, 114
information processing, xiv
informational landscape, 243
Innate immunity, 76
intelligence, 119
Interaction/Interactive Machines, 176
interpretation, 38

interpreter, 24
Intuition, xix
irreversibility, 44

junk DNA, 55

Klein bottle, 179
knowledge, 119

la Langue, 21
languaging, 21, 192
levels of organization, 31
life, 167
Ligand-receptor binding, 84
linguistics, 20
Lymphocytes, 79

Map, 6, 61
Mathematics, 20
Maxwell's Demon, 177
measurement, 177, 179
mechanical, xiii
mediated activity, 36
memorization, 229
memory, 77
mesoscopic, 169
meta-language, 21, 55
metabolism, 222
metaphor, 131
Methylation, 69, 70
Mind, 246
Möbius strip, 182
Multidisciplinarity, xvii

ncRNAs, 60
neo-vitalism, xx
non-covalent, 23

oblivion, 44
observer, 24, 246

paradox, 26
Parole, 21
perturbations, 109
physics of computation, 41
Plasticity, 187
Poetry, 261
poiesis, 85
polysemy, 39, 67, 138
post-modernist, 39
pragmatic, 20
projection, 242
Proteins, 24

quantum coin-flip gate, 205
quantum computing, 68, 202
qubit, 205

Recursive-hierarchy, 49, 86, 215
reductionism, xii, 5
repetition, 28, 244
RNA interference, 193
Russell's Paradox, 27

Secondness, 25
self and non-self, 95
self-determination, 220
self-reference, 27
self-referential, 68
selfish genes, 7
semantic shift, 172
semantics, 20
semiosis, 44, 111
semiotic, 42
Sign, 46, 61
sign-mediated, 36
silence, 70
Simpson's Paradox, 246
Singularity, 241
soft matter, 263
specificity enigma, 83
square root of NOT, 207
superposition, 67
symmetry, 88
symmetry restoration, 91, 182
synsymmetry, 87
Syntax, 20

tardigrade, 215
the book of life, 15

the central dogma, 36
The First Law of Human Perception, 250
The immune system, 75
the lock-and-key metaphor, 43, 83
the linguistic metaphor, 131
Theory of Types, 26
Thirdness, 25
thought collective, 6
thought style, 13
Tolerance, 102, 122
transgradience, 138, 140, 239
translation, 37
transmutation, 43
Turing machine, 28

utterance, 162

wound healing, 187

www.ingramcontent.com/pod-product-compliance
Lightning Source LLC
LaVergne TN
LVHW011914200926
848484LV00002B/623